SCIENCE QUIZ BOOK

1100 Questions and Answers

Rajeev Garg
M.Sc., M.Tech.

Amit Garg
M.Sc., M.C.A., Ph.D

Published by:

V&S PUBLISHERS

F-2/16, Ansari road, Daryaganj, New Delhi-110002
☎ 23240026, 23240027 • *Fax:* 011-23240028
Email: info@vspublishers.com • *Website:* www.vspublishers.com
Online Brandstore: *amazon.in/vspublishers*

Regional Office : Hyderabad
5-1-707/1, Brij Bhawan (Beside Central Bank of India Lane)
Bank Street, Koti, Hyderabad - 500 095
☎ 040-24737290
E-mail: vspublishershyd@gmail.com

Branch Office : Mumbai
Jaywant Industrial Estate, 1st Floor–108, Tardeo Road
Opposite Sobo Central Mall, Mumbai – 400 034
☎ 022-23510736
E-mail: vspublishersmum@gmail.com

BUY OUR BOOKS FROM: AMAZON FLIPKART

© **Copyright:** *V&S PUBLISHERS*
ISBN 978-93-81843-1-2
Edition 2021

DISCLAIMER

While every attempt has been made to provide accurate and timely information in this book, neither the author nor the publisher assumes any responsibility for errors, unintended omissions or commissions detected therein. The author and publisher make no representation or warranty with respect to the comprehensiveness or completeness of the contents provided.

All matters included have been simplified under professional guidance for general information only without any warranty for applicability on an individual. Any mention of an organization or a website in the book by way of citation or as a source of additional information doesn't imply the endorsement of the content either by the author or the publisher. It is possible that websites cited may have changed or removed between the time of editing and publishing the book.

Results from using the expert opinion in this book will be totally dependent on individual circumstances and factors beyond the control of the author and the publisher.

It makes sense to elicit advice from well informed sources before implementing the ideas given in the book. The reader assumes full responsibility for the consequences arising out from reading this book. For proper guidance, it is advisable to read the book under the watchful eyes of parents/guardian. The purchaser of this book assumes all responsibility for the use of given materials and information. The copyright of the entire content of this book rests with the author/publisher. Any infringement/ transmission of the cover design, text or illustrations, in any form, by any means, by any entity will invite legal action and be responsible for consequences thereon.

Printed at : Param Offsetters, Okhla, New Delhi–110020

Preface

India is a vast country where millions of candidates appear in X and XII class examinations, in various competitions and interviews every year. Almost in all the examinations short questions on various science subjects are asked. Students search many books for the preparation of these examinations. There is a great need of a book in short question-answer form dealing with various science subjects to guide the students properly. With this aim in view the present *Science Quiz Book* has been written. The overwhelming response to our earlier editions is what set us about the task of improving and updating the volume as a revised form in your hands.

The revised book contains about 1100 questions and answers on different topics of science. It has been divided into 29 chapters. Nine chapters have been devoted on the topics of modern science such as Computers, Communications, Robotics, Masers and Lasers, Energy, Space Exploration, Nuclear Sciences etc. Five chapters deal with biological science such as Plants Kingdom, Animal World, Human Body, Human Diseases, Medicine and Medical Engineering. One full chapter deals with Chemistry. Apart from these, several chapters have been devoted to Universe, Science Laws, Scientific Instruments, Domestic Appliances, Everyday Science etc. One chapter exclusively deals with the scientific achievements of India so that the reader may get the first-hand knowledge of the scientific achievements of the country. A very important chapter has also been given which presents the recent scientific achievements of the world like NMR Scanner, Acoustic Microscope, Currency Counting Machine and other revolutions. The book has been profusely illustrated. These illustrations give a better understanding of the subject matter. The language used is quite comprehensive so that the reader may not feel any difficulty in understanding the scientific concepts. No mathematical formulae have been used in the book because they make the subject matter difficult and dry.

We hope the present edition of the book will be more useful for the secondary students and for those who are appearing in various competitive examinations and interviews. It will also be useful for the students of higher classes and general readers.

—**Publishers**

Contents

1. Important Branches of Science 7
2. Computers 11
3. Electronics 15
4. Communications 21
5. Robots 26
6. Maser and Laser 29
7. Energy 31
8. Modern Technology 39
9. Nuclear Sciences 48
10. Space Exploration 56
11. Time 64
12. Chemistry 69
13. Universe 84
14. Plant Kingdom 94
15. Animal World 100
16. Human Body 117
17. Human Diseases 132
18. Medicine and Medical Engineering 138
19. Units 144
20. Science Laws and Effects 148
21. Inventions and Inventors 153
22. Important Scientific Instruments 158
23. Scientific Achievements of India 164
24. Famous Scientists of India 168
25. Domestic Appliances 174
26. Everyday Science 178
27. Miscellaneous 184
28. Recent Scientific Achievements 192
29. Scientific Abbreviations 201

1. IMPORTANT BRANCHES OF SCIENCE

Which branch of Science deals with the study of plants?
Botany.

Which branch of Botany deals with the classification and identification of plants?
Taxonomy.

Which branch deals with the study of animals?
Zoology.

Which branch of zoology deals with the shells of molluscs?
Conchology.

What is cryogenics?
Branch of physics which deals with the production, control and application of very low temperatures.

Which branch of biology deals with the study of cells?
Cytology.

What is bioengineering?
It is the application of engineering to biology and medicine.

Name the art in which acrobatic feats are performed?
Acrobatics. It refers to the performance of difficult physical acts. (Fig. 1.1)

Fig. 1.1 Acrobatics

What is ecology?
Ecology is a branch of biology. It is the study of the relationship among organisms and environment in which they live, including all living and non-living components.

What is the science of generation?
Genesiology.

What is heliotherapy?
Treatment of disease by sunlight. (Fig. 1.2)

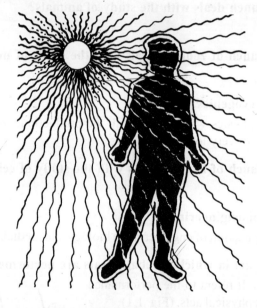

Fig. 1.2 Suncure

Name the branch of science which deals with the study of teeth.
Odontology.

What is phonetics?
Phonetics is the study of speech sounds and their production, transmission, reception, etc.

What is orthopaedics?
Orthopaedics is the science of prevention, diagnosis and treatment of diseases and abnormalities of the musculo-skeletal system. (Fig. 1.3)

What is pharmacology?
Pharmacology is the study of properties of drugs and their effects on the human body. It is a branch of medical science.

Which branch of science deals with the study of tuberculosis of lungs?
Phthisiology.

What is phrenology?
It is the study of the faculties and qualities of mind from the shape of skull.

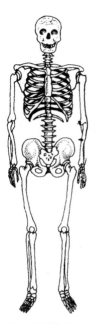

Fig. 1.3 Skeleton

What is gynaecology?
Gynaecology is a specialised branch of medical science. It deals with scientific study and treatment of diseases and disorders of the female reproductive system.

Which branch of science deals with the study of rocks, their mineral compositions and their origins?
Petrology.

What is volcanology?
It is the study of volcanoes and the geological phenomena that cause them.

Name the science which deals with fruits and fruit growings.
Pomology.

What is radiology?
Radiology is the study of X-rays and radioactivity. It is very useful in detecting cancer and other diseases.

Which branch of science deals with nature, origin and movement of moon?
Selenology.

What is the study of poisons called?
Toxicology.

What is teleology?
Teleology is the study of evidence in nature.

What is palaebotany?
The study of ancient plants by means of their remains found as fossils in rocks.

What is materia medica?
A branch of medical science concerned with the preparation and prescribing of medications and drugs.

What is meteorology?
It is the study of the atmosphere.

What is seismology?
Seismology is a branch of earth science. It is the study of earthquakes and how their shock waves travel through the earth.

What is the science of law called?
Jurisprudence.

What is the science of dendrology?
Science related to the study of trees and shrubs.

What is Karyology?
It is related to the study of nucleus.

●●●

2. COMPUTERS

What is a computer?

A computer is a programmable electronic machine used for performing calculations and other symbol manipulation tasks, quickly and accurately. They are being widely used in science, engineering, medicine, space, games, communications etc.

What are the essential parts of a computer?

Basic components of a computer is the CPU (Central Processing Unit), which performs all the computations. This is supported by memory which holds data and the current program and "logic arrays" which move information around the system. (Fig. 2.1)

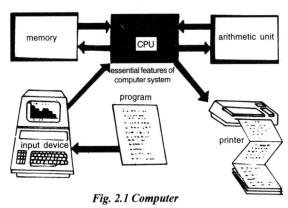

Fig. 2.1 Computer

How many types of computers are there?

There are mainly three types of computers: the analog computer which measures one quantity in terms of another, the digital computer which solves problems by using numbers, and the hybrid computer which has components of both analogue and digital computers.

When was the first digital computer developed and by whom?

In 1944, Professor Howard H. Aiken of the Harvard University and International Business Machines (IBM) developed a digital computer. The first true electronic computer was ENIAC (Electronic Numeric Integrator and Calculator), built in 1946 by engineers at the University of Pennsylvania. In 1949, John Von Neumann's computer EDVAC (Electronic Discrete Variable Computer) was the first to use binary arithmetic and to store its operating instructions internally. It still forms the basis for today's computers.

Which arithmetic is used by a computer?

A computer uses binary code in which numbers and alphabets are converted in the digits of one and zero.

What is the function of memory unit?

In the memory unit the data of the given problem and the programme to solve it are kept separately. Internal memory unit has metal oxide semiconductor field effect transistor and the external memory unit has a magnetic tape, the magnetic drum and a magnetic disc.

How does a digital computer work?

A digital computer is fed with all the information with a programme of instructions on what to do with it. Programmed data is channelled into the central processing unit by an input machine. Programmed data goes into the storage unit or core store. Calculations are then performed by the arithmetic and logic unit. All storage and calculation operations are controlled by the control unit. Answers are given by an output machine. (Fig. 2.2).

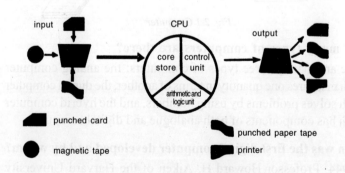

Fig. 2.2 Computer functions

How are computer generations classified?

These are classified under five broad groups:
(a) First generation developed in the 1940s and 1950s, made from valves and wire circuits.
(b) Second generation from early 1960s based on transistors and printed circuits.
(c) Third generation from the late 1960s using integrated circuits.
(d) Fourth generation using micro processors and sophisticated programme languages still in use in the 1990s.
(e) Fifth generation, based on parallel processing very large scale integration.

What is meant by the computer programming languages?

Computers are designed to understand certain languages. Some of the important languages are: BASIC (Beginner's All-purpose Symbolic Instructions Code), COBOL (Common Business Oriented Language), FORTRAN (Formula Translation), LISP (List Processing), PASCAL (French acronym), widely used for teaching programming in colleges/universities, PROLOG (Programming in Logic), etc.

What do you mean by the of terms 'hardware' and 'software'?

All the devices, circuitry and whatever we can touch and feel in a computer system is known as hardware. Software, in general, refers to the groups of instructions given to the computer to make it perform certain operations.

What are the important applications of computers?

Computers are being applied in science, engineering and industry, medicine, business management, telecommunications, education, transportation, banks, space research, etc. One can play games like chess and poker.

What are computer graphics?

Use of computers to display and manipulate information in pictorial form by scanning an image or by drawing with a mouse on graphics tablet.

How does a computer function like a man?

The input unit of a computer is equivalent to our eyes and ears. The central processing unit is equivalent to the brain. The output unit is equivalent to our hands and mouth. (Fig. 2.3 and 2.4)

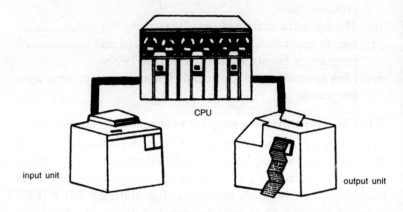

Fig. 2.3 Comparison of a computer and man

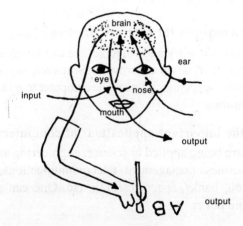

Fig. 2.4 Man

●●●

3. ELECTRONICS

What is electronics?
The branch of physics dealing with the production and control of electrons in devices such as semiconductors, vacuum tubes and various instruments.

What is a thermionic valve?
Electron tubes in which electrons are emitted from a heated cathode. The British physicist John Ambrose Fleming developed the first valve in 1904 for use in radio receivers. It was a diode valve containing a cathode and anode enclosed in an evacuated glass bulb. Two years later the American physicist Lee de Forest produced a triode valve which has a third electrode called grid put between anode and cathode. (Fig. 3.1)

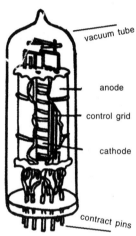

Fig. 3.1 Thermionic valve

How does a diode work and what is its use?
When the cathode is heated, electrons are emitted. These are attracted towards the positively charged anode by which the current flows through the valve but only in one direction towards the positive anode. If the anode becomes negatively charged, no current would flow. Diode valve is used for converting AC into DC in rectifiers. (Fig. 3.2)

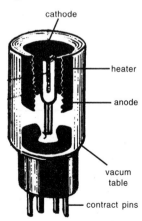

Fig. 3.2 Diode valve

How does a microwave oven work?

Microwave ovens cook food by means of radiation. This radiation is generated by a tube called magnetron.

What is a semiconductor?

A material whose electrical conductivity lies between those of a conductor and insulator is known as a semiconductor. The group of elements known as metalloids are semi-conductors such as silicon, germanium, tellurium and selenium.

What is a p-n junction semiconductor?

A p-n junction semiconductor is just like a diode valve but it is made of silicon or germanium. When an element like phosphorus is doped as impurity in silicon it becomes a n-type semiconductor. When another impurity like boron or aluminium is doped in silicon it becomes a p-type semiconductor. By putting a n-type and p-type semiconductor together, it becomes a p-n junction semiconductor. In such a semiconductor, current flows only in one direction due to the recombination of electrons and holes. A hole is where an electron is missing. It has positive charge. (Fig. 3.3)

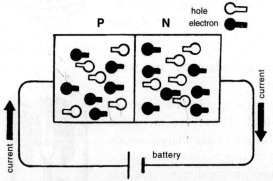

Fig. 3.3 Operation of the p-n junction semiconductor

What is a thyristor?

It is a type of rectifier of an electronic device that conducts electricity in one direction. Thyristor is composed of layers of semiconductor material sandwiched between two electrons called anode and cathode.

How does a 'n-p-n' transistor work?

A transistor is just equivalent to a triode valve. In a 'n-p-n' transistor, the voltage of the base must be more positive than the voltage of the emitter. The voltage of the collector must be more positive than the voltage of the base. In this way, the negatively charged electrons move from the emitter, through the base to the collector. The flow of electrons produces a current. The number of electrons in the base controls the flow of electrons from the emitter to the collector.

How does an electric razor work?

An electric mechanism sets in motion one or more razor blades. Above the blade there is a razor metal head. The facial hair penetrates this head and is cut by the moving blade.

What is an amplifier?

An amplifier is an electronic device that increases the strength or power of an electrical signal. It consists of vacuum tubes or transistors connected together in a circuit. Amplifiers are used to produce high quality sound.

How does a 'p-n-p' transistor work?

In a 'p-n-p' transistor, it is the positively charged holes that move from the emitter to collector. In this, the voltage of base is more negative than the voltage of emitter and also collector is more negative than the base. (Fig. 3.4)

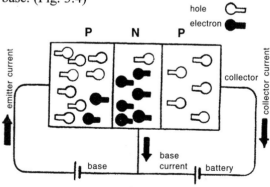

Fig. 3.4 Operation of the transistor

What is an oscillator?

An oscillator is an electronic device which produces electrical signals of desired frequency. In fact, it is a feedback amplifier that strengthens a signal and then feeds part of the amplified signal back into itself to make its own output. They are used in radio and television receivers and in other equipments.

What is an oscilloscope?

An oscilloscope is an electronic instrument which contains a cathode ray tube similar to the tube found in television sets. It contains a screen, an electron gun, anodes, focussing plates, etc. Readings are displayed graphically on the screen of a cathode ray tube. (Fig. 3.5)

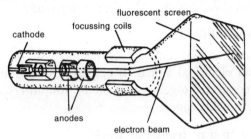

Fig. 3.5 Cathode ray tube

What is an antenna?

An antenna is a piece of equipment used for sending and receiving electrical messages. It is a basic part of all electronic communication systems. It is used for radio, television, radar and radio telescope operations. Different types of antenna are shown in fig. 3.6.

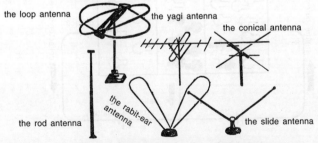

Fig. 3.6 Different types of receiving antennas

What is a microphone?

A microphone is a device which converts sound waves into electric current. The essential parts of a microphone are a diaphragm and a device which converts the vibrations of a diaphragm into variable electric current. The device may be a moving coil, a piezoelectric crystal, a capacitor or carbon granules.

What is an integrated circuit?

An ordinary circuit consists of various components connected together with wires. An integrated circuit has all its components on one small 'chip' of silicon and different impurities are added to different parts of a silicon chip. This makes the different parts behave like resistors, diodes or triodes.

What is a teleprinter?

It is a telegraph transmitter with the help of which we can send messages at the rate of more than 50 words per minute. Signals are sent by pressing the keys of a machine which resembles a typewriter.

How does a loudspeaker work?

A loudspeaker is a device that converts electrical signals into sound waves. A common loudspeaker is a moving coil speaker as shown in fig. 3.7. A coil of wire is attached to the apex of large paper cone

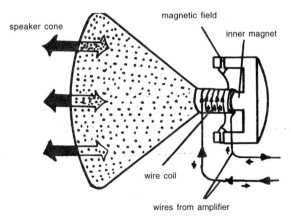

Fig. 3.7 Loudspeaker

and placed between the poles of a permanent magnet. When an electrical current passes through the coil, it sets up a magnetic field. This produces vibrations in the cone by which sound is emitted.

What is a tape-recorder?

A tape-recorder is a machine that can record and playback sound. It makes use of a magnetic tape for recording and reproducing sound.

How does a tape-recorder work?

A tape-recorder is an electronic machine that can record sound on a tape coated with iron oxide. In this device first the sound is converted into electric current by a microphone. The current is amplified by an amplifier. The current is then recorded on a magnetic tape in the form of magnetic field. The tape is run by an electric motor. To reproduce the sound, the magnetic field again changes it into sound. (Fig. 3.8)

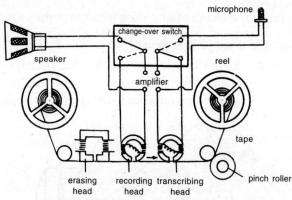

Fig. 3.8 Mechanism of the tape-recorder

What is a rectifier?

It is an electronic device that allows electricity to pass in one direction only. It converts A.C. into D.C.

●●●

4. COMMUNICATIONS

What were the primitive methods of communication?
Primitive man communicated over a distance by shouting or blowing a horn, by creating smoke or fire and by beating drums.

What are the modern means of communication?
Telegraph, teleprinter, telephone, radio, television, telex, fax, pager, internet, e-mail, cinema, newspapers, mobile phones etc., are the modern means of communication.

What is internet?
Internet is the world's largest computer network. The network is basically a bunch of computers connected together. Actually the internet is not really a network, it is a network of networks — all freely exchanging information.
It has become the fastest and most reliable way to move or exchange information.

What is a telegraph?
The telegraph is a means of sending messages by electricity through wires.

What is a Morse code?
Samuel Morse developed a code in which alphabets could be transmitted as dots and dashes. A pause is given between the letters. Messages in Morse Code could be sent from ships during day or night using flashing lights or whistles.

What is a semaphore?
Semaphore is a system of signalling by changing the position of arms, a flag or light. The semaphore system of communication is very useful for signalling between the ships.

What are the main parts of telegraph key?

The main parts are backstop, frame, hammer, anvil, circuit closer lever, key lever, etc. (Fig. 4.1)

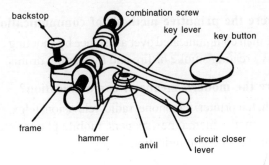

Fig. 4.1 Telegraph key

How does a telephone work?

A telephone has two main parts: the mouthpiece and the earpiece. Both are enclosed in one case. When we speak into the mouthpiece, a diaphragm attached to it starts vibrating and in accordance to these vibrations a varying current is produced. The current is carried by the telephone line wire to the receiver of another telephone.

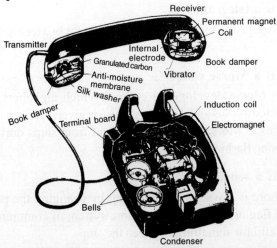

Fig. 4.2 Mechanism of a telephone transmitter and receiver

This varying current is converted into sound in the receiver. Thus our voice reaches to the other end and vice-versa. (Fig. 4.2)

What are the various types of telephones?
Today we have cordless telephones, portable telephones, video-telephones and cellular or mobile phones.

What is a video telephone?
It is a communication device used for simultaneous exchange of visual images and associated speech. In this type of telephone one can see the picture of the person talking at the other end.

What is a cellular phone?
It is a mobile radio telephone — a network connected to telephone system by computer-controlled communication system.

Which company of the world manufactures the maximum mobile telephones?
NOKIA.

What is multimedia?
Multimedia is the result of a co-ordinated work of video, audio and graphics in computer. A multimedia computer opens up a world of education, information, entertainment and so on. All multimedia programs include six common elements — texts, pictures, movies, animation, sound and increased control.

What are radio waves?
Radio waves are electromagnetic waves having wavelengths from 50 cm. to 30 metres.

What is FM transmission?
Frequency modulation (FM) is a method by which radio waves are altered for transmission of broadcasting signals. It is constant in amplitude and varies the frequency of the carrier wave in accordance with the signal being transmitted. Its advantage over AM (amplitude modulation) is its better signal-to-noise ratio.

What is sonar and how does it work?

Sonar is used to locate under water objects such as submarines, torpedos or shoals of fish. It is an acronym for Sound Navigation And Ranging. It makes use of ultrasonic waves. The transducer sends out ultrasonic waves which get reflected from the object. Time taken for any returning echoes is measured and by using velocity of sound in water, distance of the object is measured. The range and depth of the object is displayed on a cathode ray tube. (Fig. 4.3)

Fig. 4.3 Sonar

What is a radar and its use?

Radar is the acronym for Radio Detection and Ranging. It is a device for locating objects in space, finding direction and navigation by means of transmitted and reflected high frequency radio waves. It allow pilots or navigators to locate objects through darkness or a fog.

What is fax?

Fax is the common name for facsimile transmission of images over a telecommunications link, usually the telephone network. When placed on a fax machine, the original image is scanned by a transmitting device and converted into coded signals, which travel via telephone lines to the receiving fax machine where an image is created that is a copy of the original. Photographs as well as letters, drawings can be sent through fax. One can also send the photos and copies of CAT scan.

What is an electronic mail?
Electronic mail (E-mail) is a telecommunication system that enables the users of a computer network to send messages to other users. Telephone wires are also used to send signals from terminal to terminal.

What for does Internet stand?
It stands for International Network.

What are the capabilities of Internet?
With the help of internet we can get information on any subject.

What is chatting?
We can have communication with our friend or relation through chatting on Internet.

What are the requirements of internet?
A computer, telephone system and an INTERNET connection.

What is Internet cable?
A cable connection to have your internet connection.

What for does LAN stand?
LOCAL AREA NETWORK. All computers of an office or establishment are connected to each other by LAN.

●●●

5. ROBOTS

What is a robot?

A robot is an automatic machine that can perform and replace many human activities. It works according to man's order. Many robots are designed to work in situations that would be dangerous to humans — for example, in defusing bombs or in space and deep-sea exploration. All actions are controlled by a computer.

What is an android robot?

Any robot having human resemblance is called an android robot. Advances in computer programming have produced robots that can see and avoid obstacles as they move around.

How many types of Industrial robots are there?

There are mainly four types of robots namely:
 (i) Pick and place robots
 (ii) Light weight electric robots
 (iii) Heavy duty robots and
 (iv) Special purpose robots

What is the heart of a robot?

A computer is the heart of a robotic device. It changes human words into electrical signals. These signals form the language of a robot. Computers also have electronic memory banks that allow robots to recall information.

How many motions does a robot possess?

There are six types of movements in almost all robots:
 (i) Radial movement
 (ii) Wrist bend
 (iii) Wrist swivel
 (iv) Vertical
 (v) Rotational and
 (vi) Gripper

First robot hand capable of imitating the gripping action of the human hand was used in a US nuclear plant in 1966.

Which countries are making use of robots?

The chief users of robotic devices are Japan, the USA, Germany, Sweden, Britain, France, Italy and Belgium. Although industrial robots were invented in 1962 by USA, Japan took them up and used them for welding, painting and controlling other machines. By 1995 Japan had about 14,000 robots, more than all the robots in the rest of the world.

Who first coined the word 'Robot'?

This word was first coined in 1920 by the Czech writer Karel Capek. The word robot from the Czech word for 'work', comes from the play *Rossum's Universal Robots* written by him in 1920.

What is the future of robotics?

As the industrialization advances the robots will definitely replace human beings. It has a bright future.

What are the essential parts of a robot?

A robot essentially consists of a hydraulic supply, servo mechanisms, switches, computer, comparator leach control, motors, arms, etc. (Fig. 5.1)

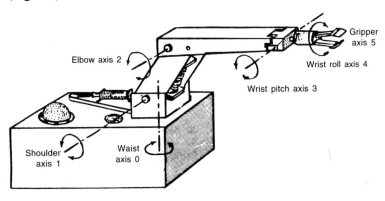

Fig. 5.1 Essential parts of a robot

What can be done by a robot?

Robots can do many works which a human being can't. For example, it can hold a red hot rod of iron and can work in radioactive environment. A robot can clean houses, pick books from library shelves, paint motor vehicles, load and unload goods. It can work in hot and poisonous environments. Some countries have developed

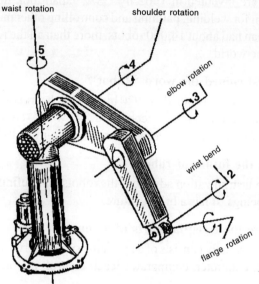

Fig. 5.2 Functions of robot

robots which can pilot even aeroplanes and ships. They can go even to the depths of oceans. They can work in factories much more efficiently than men (Fig. 5.2)

Which country is using maximum number of robots?
Japan is using maximum number of robots.

Which motor company in India is using robots?
Maruti Udyog Ltd., Gurgaon, for making cars.

What is the use of optical fibres?
Optical fibres are used in research and communications.

●●●

6. MASER AND LASER

What is a MASER?

The word maser is the short form for Microwave Amplification by Stimulated Emission of Radiation. It is a device that produces microwaves with just one frequency. It is used as a very accurate timing device. It is used in atomic clocks. It can be accurate to one second in 100,000 years. Masers are also used in astronomy to amplify radio signals coming from stars and planets. The principle has been extended to other parts of the electromagnetic spectrum as, for example, in the laser. The first maser was made from ammonia gas.

Who invented maser?

The maser was invented by an American Physicist Charles H. Townes in 1955.

On which fact is the action of maser based?

Maser action is based on the fact that irradication at the frequency concerned stimulates the process. If more atoms are in the higher energy (exited) state than in the lower state — incident waves cause more emission than absorption resulting in amplification of the original wave.

Who made the first LASER?

Theodore H. Haiman of the U.S.A. made the first LASER in June 1960. It was a ruby laser capable of giving pulses of bright red light — having a wavelength of 6943 A°. (Fig. 6.1)

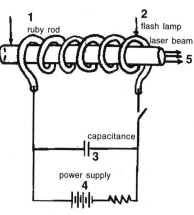

Fig. 6.1 Schematic presentation of a ruby laser

What are the uses of laser beams in modern technology?
Uses of laser beams include communications, cutting, drilling, welding, satellite tracking, medical and biological research, surgery, and entertainment like light shows etc.

What are the medical uses of laser?
They are being used for welding the detached retina. They are also being used in the treatment of cataract and glaucoma. They are useful in bloodless surgery and in cancer treatment.

What is a beam weapon?
A beam weapon is capable of destroying a target by means of high energy laser beams. Most frequently discussed are the two types: (i) High Energy Laser (HEL), and (ii) Charged Particle Beam (CPB). The HEL produces a beam of high accuracy, which burns through the surface of its targets. The CPB uses either electrons or protons to slice through its target.

When was a laser used to throw light at the moon's surface?
In 1969, a ruby laser was used to throw light at the moon's surface by the Massachusetts Institute of Technology and a reflection was detected $2^{1}/_{2}$ seconds later. A retro-reflector was mounted on the surface of moon in Apollo flight.

What are defence applications of lasers?
Lasers are being used as range finders, target seekers and for pin-point dropping of bombs.

What are the other uses of lasers?
Lasers are being used in printers, fingerprint detection, television, pollution detection, weather forecasting etc.

What is a hologram?
A hologram is a three-dimensional record of an object made with a laser and viewed by a laser.

What are different lasers?
Ruby laser, helium-neon laser, carbon-dioxide laser, Nitrogen laser, YAG laser, glass laser are different types of lasers being used in different applications.

●●●

7. ENERGY

What is energy?

The capacity of doing work is called energy. When a force moves an object, work is done. When a boy lifts a book from the ground, the boy does work and thus some energy is used.

What are the different forms of energy?

Energy can exist mainly in eight forms. These are: mechanical energy, light energy, sound energy, heat energy, magnetic energy, electrical energy, chemical and atomic energy.

What are renewable resources of energy?

The sources of energy which are continuously produced in nature are called renewable sources of energy. They include plant, food and feed, wood, fuel, falling water, geothermal power, etc. Renewable sources depend ultimately on sun's energy.

What are non-renewable resources of energy?

These are those sources of energy which have been accumulated over the ages and are not quickly replaceable when they are exhausted. Non-renewable resources include the fossil fuels like coal, oil and gas and nuclear fission 'fuels' for example, uranium-235.

What are the two forms of mechanical energy?

The two forms of mechanical energy are kinetic energy and potential energy. Kinetic energy is the energy of motion and potential energy is the energy of position. A running train or fan possess kinetic energy while a soap bubble or spring possess potential energy.

What are the reliable sources of energy for future?

The sun and nuclear fusion are the reliable sources of energy for future. Controlled fusion is however, long way off.

What is an electric-power station?

An electric-power station is a place where electricity is generated. Most common kinds of power plants are steam-turbine plants, hydroelectric plants and atomic or nuclear plants.

What are the major sources of energy?

Coal, oil, water, sun, wind, garbage, internal heat of the earth, atomic nucleus are the main sources of energy. Different energy percentages are shown in fig. 7.1.

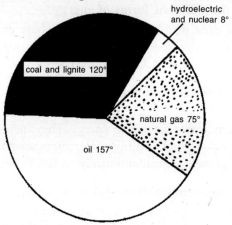

Fig. 7.1 Pie chart: Energy sources of the world

How does a steam-turbine power-plant generate electricity?

A steam-turbine power-plant uses oil, coal or gas to produce steam by heating water. They are burnt to boil water and produce steam. This steam spins the wheels of a turbine which in turn run the generator. This generator produces electricity.

What are the main energy sources of electricity generation?

They are coal, oil, water power (hydroelectricity) natural gas and nuclear power, with limited contributions from wind power, tidal power and geothermal power.

What is heat?

Heat is a form of energy produced by the motion of molecules.

How does a hydroelectric power-plant generate electricity?

The falling water from dam runs a turbine which in turn runs the generator. This generator produces electricity. In a typical hydroelectric power station, water is stored in a reservoir and is piped into water turbines connected to electricity generators. (Fig. 7.2)

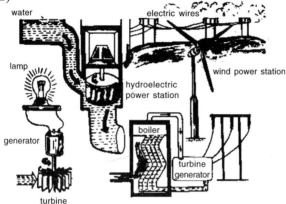

Fig. 7.2 Different power stations

How does a steam engine work?

In a steam engine, the super heated steam is made to pass into the steam chest. The steam is then fed to the cylinders, where it forces the pistons back and forth alternately. The motion of the piston is transferred to a fly wheel with the help of a crankshaft, by which the train moves. (Fig. 7.3)

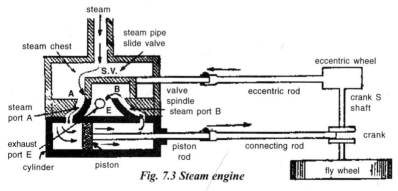

Fig. 7.3 Steam engine

How does a petrol engine work?

A petrol engine is an internal combustion engine in which petrol burns inside the engine and heat energy of the fuel gets converted into mechanical energy. It is a four-stroke engine. In the first stroke petrol vapours enter into the cylinder. In the second stroke fuel is burnt by an electric spark. In the third stroke piston moves, and in the fourth stroke unused gases come out. The movement of the piston turns the wheel and as a result the vehicle moves. (Fig. 7.4)

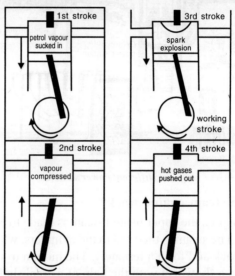

Fig. 7.4 Petrol engine

What is a photon?

Photon in physics is the smallest package or particle, or quantum of energy in which light or any other form of electro-magnetic radiation is emitted. The energy associated with a photon is $h\nu$.

What is the fundamental particle of light?

Photon is the fundamental particle of light.

What is a solar cell?

A solar cell is a device that converts light energy into electrical energy. Solar cells are usually made from silicon, which generate electricity when illuminated by sunlight.

What is a photocell?

A photocell is a device which converts light energy into electrical energy. It is also used for measuring or detecting light or other electromagnetic radiation since its electrical state is altered by the effect of light. It makes use of photosensitive material such as silicon, potassium, etc. to convert light into electrical energy. A photo transistor is shown in fig. 7.5.

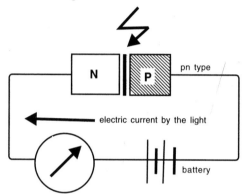

Fig. 7.5 The photo transistor detects light

How does a dry cell produce electrical energy?

A dry cell consists of a cylindrical vessel of zinc which acts as negative electrode. A carbon rod is put at the centre of this container which works as positive electrode. A paste of ammonium chloride and manganese-dioxide is filled in the container to increase the life of the cell. The chemical energy gets converted into electrical energy in the cell. The common dry-cell is a primary-cell battery and it is dangerous to try to recharge it.

What are the nickle-cadmium batteries?

Cadmium is a soft silver-white material used in batteries. It occurs in nature as a sulfide in zinc ores. Introduction of rechargeable nickel-cadmium batteries have revolutionized portable electronic news gathering (sound recording, video) and information processing (computing). They are stable short-term source of power and noise-proof.

What is telephoto-transmission?

The transmission of photographs and text by wire or radio is called telephoto-transmision. The image is divided into lines and a phototube changes these lines into electric current. The current is then carried over telephone cable. On the receiving end, a sensitive paper changes it into photo again. (Fig. 7.6)

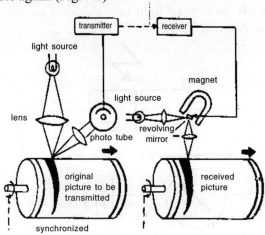

Fig. 7.6 Telephoto transmission

How does a battery produce electrical energy?

A battery is a device that produces electricity by chemical reaction. A lead-acid battery consists of two electrodes, one made of lead and the other of lead-oxide. These two electrodes are dipped in dilute sulphuric acid. At the positive electrode, lead dioxide reacts with sulphuric acid and obtains two electrons from wire, while at the negative electrode lead reacts with sulphate ions and provides two electrons to the wire. The net effect is that electrons start flowing from negative plate to positive plate. These batteries are rechargeable.

What is the difference between a light bulb and a fluorescent lamp?

An electric bulb produces light by the heating effect of electricity, *i.e.,* electric current heats up the tungsten filament, but the fluorescent tube works on the principle of electric discharge through gases.

How does mercury lamp produce light?

A mercury lamp consists of mercury vapours at high pressure filled in an evacuated glass tube. When electricity passes through high pressure mercury vapours, light is emitted due to electrical discharge. (Fig. 7.7)

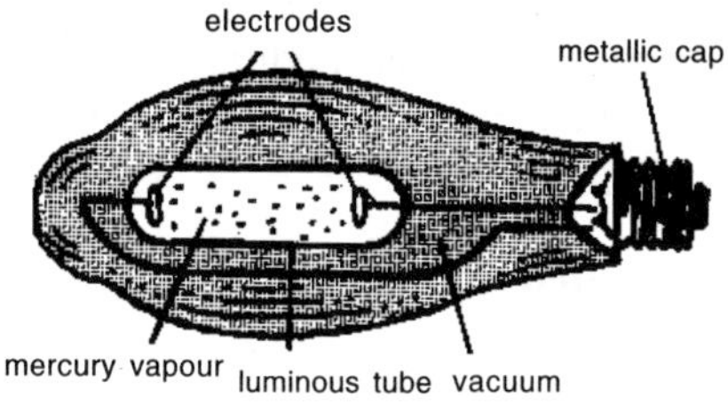

Fig. 7.7 High pressure mercury lamp

How does a sodium lamp work?

A sodium lamp is a source of yellow light. It produces light due to electrical discharge of sodium vapour. Different parts of sodium lamp are shown in fig. 7.8.

Why are electric wires covered with plastic or rubber?

The plastic or rubber covering insulates the wire, *i.e.,* it prevents the leaking of electricity through the sides of wire. An insulated wire is safe to touch but a bare wire gives a shock on touching.

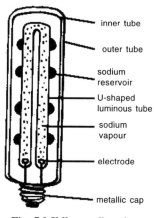

Fig. 7.8 Yellow sodium lamp

What colour of light is most readily absorbed by water?

Red colour is absorbed the most by water and green is absorbed the least.

What are infrared rays?

Rays having wavelengths higher than that of red light (7500A°) are called infrared rays. These rays produce heat. Infrared lamps are used for treating the muscle and joint pains. These rays are also used for guiding and controlling missiles. Infrared radiation is also used in medical photography and treatment and in industry, astronomy and criminology.

How does sound travel from one place to another?

Sound travels from one place to another in the form of longitudinal waves. All sound waves in air travel with a speed dependent on temperature. It needs a material medium to travel, sound waves can travel through solids, liquids and gases. Sound travels fastest in solids and then in liquids.

How is sound recorded and reproduced in a film?

For recording the sound in a film, a microphone converts sound waves into electric current. This changing current changes the light output of the lamp. This changed light is recorded in the film (Fig. 7.9). Optical recordings used for motion-picture sound tracks, converts the microphone signals into a photographic exposure in the film using a light beam and a variable shutter. The sound is played back by shining a light beam through the track onto a photoelectric cell. This reproduces electrical signals, which are amplified and fed into a loudspeaker. (Fig. 7.10)

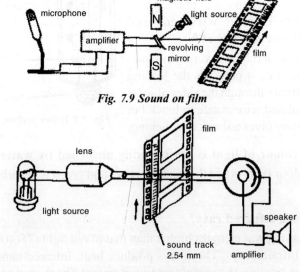

Fig. 7.9 Sound on film

Fig. 7.10 Reproduction of sound

●●●

8. MODERN TECHNOLOGY

What is annealing?
Annealing is a type of heat treatment in which metal or glass is heated and then allowed to cool slowly in given time, at a given temperature. By annealing, internal stresses are removed.

What is anodizing?
Anodizing is a process of increasing the corrosion resistance of certain metals like aluminium by building up a protective oxide layer on the surface. The natural corrosion resistance is provided by a thin film of aluminium oxide. The process is called anodizing because the metal is made the anode in an electrolytic cell.

What is an antifreeze?
An antifreeze is a chemical compound added to water particularly in a car radiator to prevent it from forming into ice. Ethylene Glycol is a well known antifreeze compound.

What is asbestos?
It is a fibrous form of certain minerals such as crocodilite and chrysolite. It is highly heat proof. Commercial asbestos is generally either made from serpentine or from sodium iron silicate.

What is bakelite?
It is a synthetic plastic made from phenol and formaldehyde. It is used as an electric insulator and as an adhesive.

What is a bolometer?
It is a sensitive instrument used for measuring radiation by registering the change in electrical resistance of a fine wire when it is exposed to heat or light.

What is a bubble-chamber?
It is a chamber usually containing liquid hydrogen under pressure and just below its boiling point. It is used in nuclear research for

showing the path of charged atomic particles. It was invented by US physicist Donald Glaser in 1952.

What for is radiosonde used?

It is a meteorological device used to study upper atmosphere. Radiosonde balloon carries a compact package of meteorological instruments and a radio transmitter which record temperature, dew point, humidity, pressure and wind speed in the upper atmosphere.

How is diamond made synthetically?

Diamond is made synthetically by heating graphite to about 3000°C under high pressure.

What is neoprene?

It is a kind of synthetic rubber developed in USA in 1931 from the polymerization of chloroprene. It is much more resistant to heat, light, oxidation and petroleum than is ordinary rubber. It is used in paints and crepe soles etc.

What is fibreglass?

Fibreglass is an artificial material composed of tiny threads of glass. It retains the tensile strength of glass while flexible. It is strong, durable and fireproof.

What is an escalator?

An escalator is a moving stairway by which passengers are carried either up or down. It is used in large stores, buildings, railroads, bus and airline terminals.

What does 'SCUBA' mean?

SCUBA is the abbreviation of "Self-contained Under-water Breathing Apparatus". It contains one or two compressed air tanks and a connecting hose with a mouthpiece.

What is a crane?

A crane is a machine, made on the principle of pulleys, which can lift and move heavy objects. Its name has originated from crane bird which has a long neck.

How are X-rays produced?

X-rays are produced in an X-ray tube. When cathode rays of high energy are bombarded on targets of metals like molybdenum or tungsten, X-rays are produced. (Fig. 8.1)

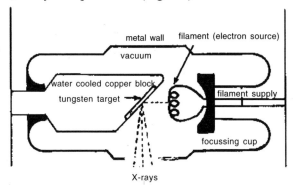

Fig. 8.1 X-ray tube

How does a submarine stay under water?

Each submarine has several large ballast tanks, which can be filled with either water or air. When they are filled with air, the submarine will float on the surface of water. When water is pumped into tanks, it starts sinking. The crew of the submarine can make their vessel rise or sink to various levels of water by regulating the amount of water and air in the tanks. (Fig. 8.2)

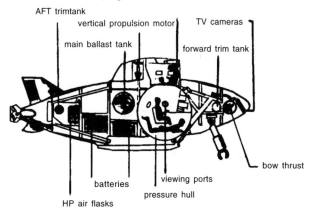

Fig. 8.2 Submarine

When was the first nuclear-powered submarine launched?
In 1954, the USA launched the first nuclear-powered submarine — the Nautilus.

What is a ram jet?
It is a simple kind of jet engine used in some guided missiles. It comes into operation only at high speed. Then air is 'rammed' into the combustion chamber, into which fuel is sprayed and ignited. (Fig. 8.3)

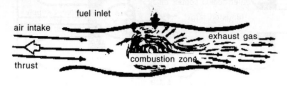

Fig. 8.3 Ram jet

What is a bathyscaphe?
A bathyscaphe is a specially designed vessel used for deep-sea exploration. It has a cabin for two persons to sit, and a tank for fuel. There are two cylinders which can be filled with water or air to take the vessel to the desired depth. It is a very convenient device for studying the sea-bed and to understand the formation of the earth's crust and learn about marine life. (Fig. 8.4)

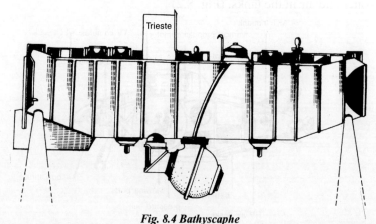

Fig. 8.4 Bathyscaphe

What is hypersonic flight?

Flights at speeds over five times the speed of sound in the air, are called hypersonic flights.

What is the speed of concord aircraft?

The concord aircraft can fly at twice the speed of sound, *i.e.,* at a speed of about 2170 km/hour or 2 mac.

Which was the first successful flying machine?

The first successful flying machine was a hot air balloon built by Joseph and Jacques Montgolfier in 1783. This first balloon attained an altitude of 1,500 feet within 10 minutes.

What are artificial or man-made fibres?

Rayon, terylene, acrilan and nylon are all man-made fibres. Rayon is made from wood. Terylene, acrilan and nylon are made from chemicals obtained from petroleum and coal.

Which is the material from which rayon is made?

Rayon is made from cellulose, which is extracted from wood pulp.

What is carbon dating?

It is a method of estimating the ages of archaeological specimens of biological origin, by measuring the radioactive carbon present in that specimen. It was developed by William F. Libby and his associates in 1947.

What is theodolite?

It is an instrument used in land surveying for measuring angles horizontally and vertically. It consists of a small telescope mounted so as to move on two graduated circles, one horizontal and other vertical, while its axes pass through the centre of the circles. (Fig. 8.5)

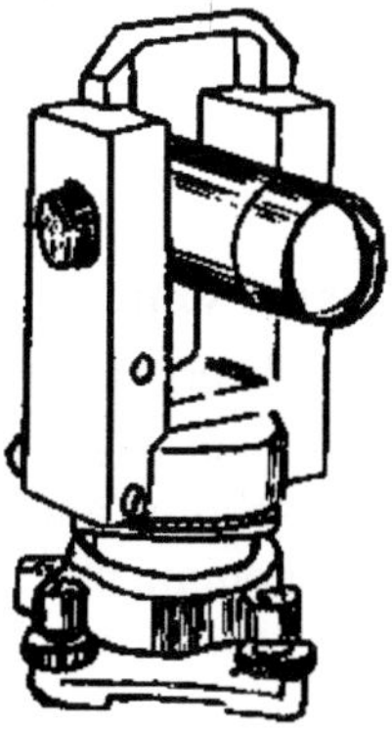

Fig. 8.5 Theodolite

How is galvanizing done?
Galvanizing of metal is the process for rendering iron rust-proof by plunging it into molten zinc *i.e.* dipping method or by electroplating it with zinc.

What is xerography?
Xerography is a process of photocopying. It was developed by America's C.F. Carlson in 1938 without the use of negatives or sensitized paper. In the process, an electrostatic image of the document to be copied is formed on a selenium coated drum. This is then dusted with ink-powder. The ink image is then transferred to a paper and fused to it by heat.

How does a camera work?
A camera is a light tight-box having a mechanism of holding photo film and a lens opposite to the film. The lens produces an inverted image of the object on the film. When shutter is pressed light enters the camera for fraction of a second and photo is recorded on the film.

What is a polaroid camera?
It is an instant picture camera invented by Edwin Land of USA in 1947. Modern polaroid cameras can produce black-and-white prints in a few seconds and colour print in less than a minute. An advanced model has automatic focussing and exposure. The film consists of layer of emulsion and colour dyes together with a pod of chemical developer. When the film is ejected the pod bursts and processing begins in the light.

Name some artificial fertilizers and mention their function.
Fertilizers supply essential elements to the soil which plants need for healthy growth. The main elements required are nitrogen, phosphorus and potassium. Some important fertilizers are urea, ammonium phosphate, potassium superphosphate etc.

How does a fire extinguisher extinguish fire?

A fire extinguisher is a portable device used to put off fire. Soda-acid extinguishers produce carbon dioxide which comes due to the reaction of sulphuric acid and sodium bicarbonate. Carbon tetrachloride is also used in fire extinguishers. A fire extinguisher is shown in fig. 8.6.

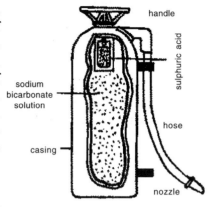

Fig. 8.6 Fire extinguisher

How are steam power stations cooled?

Steam power stations are cooled by a huge cooling tower. Water is trickled down inside the tower and is cooled as some of the water evaporates.

Where are delta-shaped wings used?

The supersonic planes and space shuttle both have delta-shaped wings. Delta is the Greek letter denoted as a triangle (Δ).

What is an excavator?

An excavator is a machine used for digging into the ground. (Fig. 8.7)

Fig. 8.7 Excavator

What are the materials used for making paper?
Wood pulp is the principal raw material used for making paper. It is obtained from spruce, pine, eucalyptus, poplar, birch and chestnut trees.

What is napalm bomb?
It is a bomb which contains highly inflammable petroleum jelly. Extensive burns are caused by napalm, because it sticks to the skin even when alight.

What is a hovercraft?
It is a vehicle which propels on a cushion of air formed by valves on the underside of its body. It can move both on land and water, a few inches above the surface. British Engineer Christopher Cockerell invented the hovercraft in 1959.

What is phytotron?
It is a big machine used for producing any type of climate. It is used for the study of environmental biology.

What is a bulldozer?
It is an earth-moving machine widely used in construction work for cleaning rocks and tree stumps and generally levelling a site. It is a kind of tractor with powerful engine and a curved blade at the front, which can be lifted and forced down by hydraulic rams. It usually has crawler, or caterpillar tracks so that it can easily move over rough grounds.

What is a mail sorting machine?
It is an automatic machine which can sort about 360 pieces of mail per minute.

How does a video tape-recorder work?
A video tape-recorder registers TV signals on a magnetic tape. It has special recording units: one for sound recording and the other for video picture recording. A TV programme recorded on video-tape can also be broadcast.

What is an automatic vending machine?

An automatic vending machine is a device which dispenses an article automatically when a coin is put in. Its source of power is electricity. Energy chart of a vending machine is shown in fig. 8.8.

How does an electric locomotive work?

An electric locomotive picks up electricity at 25,000 volts AC from overhead power lines which is converted into DC by rectifiers. This drives motors and drive wheels.

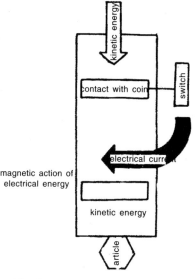

Fig. 8.8 Automatic vending machine

What is a dish washer?

It is an electrically operated machine used to clean dirty crockery and cutlery. A detergent is used to clean grease and grime.

How does a washing machine work?

A washing machine is an electrically operated machine used to clean the clothes by using detergent and water.

How does an electric iron work?

An iron is an electrically operated machine used to smooth the wrinkles of clothes. A steam iron makes steam to make the cloth slightly damp.

How does a diesel locomotive work?

A diesel engine uses in a locomotive. Diesel engine was invented diesel by Rudolf Diesel of Germany.

How does a diesel-electric locomotive work?

In this engine diesel drives an electric generator which produces electricity to drive electric motors of the engine.

●●●

9. NUCLEAR SCIENCES

What is nuclear physics?
Nuclear physics is the study of atomic nucleus. Every atom has a nucleus at its centre which is mainly composed of protons and neutrons. In nuclear physics we study the structure of nucleus, nuclear forces, nuclear reactions and energy, nuclear reactors and radioactive decay etc.

How do we get nuclear power?
Nuclear power is obtained from the energy of the nuclei of atoms. It is obtained by fission and fusion.

On what principle is atomic energy produced?
Atomic energy is produced on the principle: mass can be converted into energy. Albert Einstein gave a relation between mass and energy ($E = mc^2$). This is known as mass energy relationship.

What is nuclear fission?
When the nuclei of certain heavy elements like uranium-235 or plutonium-239 are bombarded with neutrons they split into two nuclei of barium and krypton. This is called nuclear fission. In this process of splitting of the nucleus a lot of energy (about 200 Mev) is released. (Fig. 9.1)

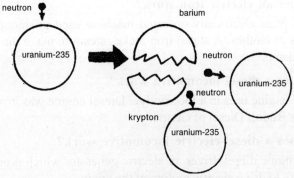

Fig. 9.1 Uranium fission

What is nuclear fusion?

Fusion is a process in which nuclei of deuterium and tritium fuse together to form helium nuclei. Hydrogen fusion is shown in figure 9.2. Fusion reactions generate much more energy than the fission reaction. Hydrogen bomb is an example of fusion reaction.

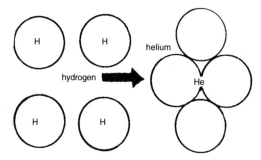

Fig. 9.2 Hydrogen fusion

What is a hydrogen bomb and which countries have developed it?

Hydrogen bomb is based on the phenomenon of nuclear fusion. In fusion process, four hydrogen nuclei combine together to form one helium nucleus. In this process a large amount of energy is liberated. Hydrogen bomb explodes with the help of an atom bomb. Two isotopes of hydrogen namely deuterium and tritium combine, and produce large amount of energy. USA, Russia, UK, France and China have developed and tested hydrogen bombs. (Fig. 9.3)

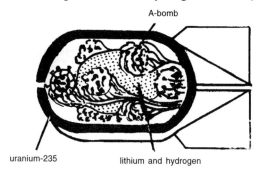

Fig. 9.3 Hydrogen bomb

When was the first atom bomb made?

The possibility was first explored in Britain from 1940, but work was transferred to USA after its entry in World War II. The first atom bomb was made in 1945 during World War II. Two atom bombs were dropped on the Japanese cities of Hiroshima and Nagasaki each on 6th August and 9th August, 1945.

How does an atom bomb work?

An atom bomb explodes due to the process of fission. Nuclei of uranium-235 are bombarded with neutrons. As soon as a nucleus splits by the bombardment of a neutron into two, three neutrons along with the energy of 200 Mev are produced. These neutrons split the other uranium nuclei and soon a chain reaction develops which produces enormous energy. This is nothing but atomic explosion. (Fig. 9.4 and Fig. 9.5)

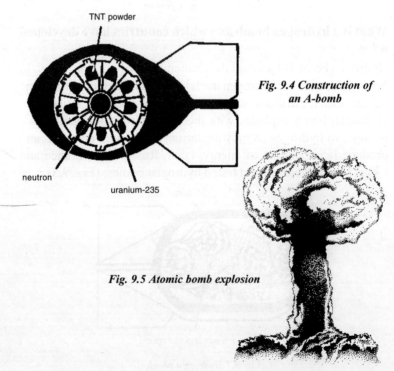

Fig. 9.4 Construction of an A-bomb

Fig. 9.5 Atomic bomb explosion

How do we express the power of an atom bomb?

The power of an atom bomb is expressed in terms of metric tons of TNT. For example, when we say a five megaton atom bomb, it means that this atom bomb will produce an energy equivalent to the energy produced by burning 50,00,000 metric tons of TNT (trinitrotoluene, *i.e.,* dynamite).

What was the name of the first atom bomb?

Little Boy, dropped on Hiroshima.

What was the name of the second atom bomb?

Fat Man (Fig. 9.6), dropped on Nagasaki.

Fig. 9.6 Nuclear bomb named Fatman

What do you mean by 'Nuclear safety measure'?

Nuclear safety measures are to avoid accidents in the operations of nuclear reactors and in the production and disposal of nuclear waste. There are however no guarantees of safety of any of the various methods of disposal.

What is a neutron bomb?

A neutron bomb is a small nuclear weapon which makes use of a neutron-rich substance for battlefield use. This emits neutrons and gamma rays, lethal within a radius of one to two kilometres. It does not destroy buildings and vegetable life but only humans and animals. Deaths are caused due to the penetration of neutrons and gamma rays.

Who discovered neutron?

Neutron was discovered in 1932 by an English physicist, Sir James Chadwick, for which he was awarded Nobel Prize for Physics in 1935.

What are elementary particles?
Elementary particles are those which cannot be divided into further simpler particles. They are basic units of matter and can decay into other smaller particles.

Name some important elementary particles.
There are three groups of elementary particles: quarks, leptons and gauge bosons. Proton, neutron, electron, mesons, neutrino, positron, photon, etc. are some of the important elementary particles.

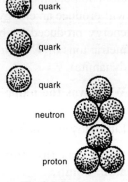

Fig. 9.7 Charges on subatomic particles

What are quarks?
Quarks are hypothetical elementary particles having slightly different mass and charge. They constitute neutrons and protons as shown in figure 9.7. To each quark there is an anti particle called antiquark.

What is radioactivity?
Some nuclei emit invisible powerful radiations. These are called radioactive isotopes and the phenomenon is called radioactivity. They emit apha, beta and gamma rays. Alpha, beta and gamma radiations are dangerous to body tissues, especially if a radioactive substance is ingested or inhaled. Alpha rays are positively charged particles, beta rays are negatively charged particles while gamma rays are electromagnetic radiations. Alpha and beta particles can be separated out by applying magnetic field as shown in fig. 9.8.

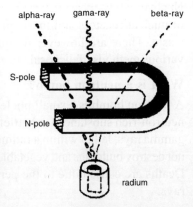

Fig. 9.8 Radioactive rays from radium

What is a particle accelerator?

A particle accelerator is a machine used to accelerate sub-atomic particles, like protons, electrons, etc. Cylotron, Van de Graaff generator, betatron, liner accelerator (linac) etc., are well known particle accelerators.

What is heavy water?

Heavy water is a compound of deuterium and oxygen. Its chemical formula is D_2O. Deuterium is an isotope of hydrogen which is heavier than ordinary hydrogen. It has been used in nuclear reactors.

Which was world's first nuclear power station producing electricity?

The first nuclear power station producing electricity was EBR-1 in the USA which started functioning on 20th December, 1951. Britain's earliest was Calder Hall (Unit I). It was opened on 27th August, 1956.

What is a fast breeder reactor?

A fast breeder reactor can be called an inexhaustible energy source. It generates electric power as well as produces fossil fuel. In this type of reactor uranium-238 is converted into plutonium-239 which can be used as nuclear fuel. Fast breeder reactors can extract about 60 times the amount of energy from uranium that thermal reactors do. In a breeder reactor sodium is used as coolant. It does not need moderator. (Fig. 9.9)

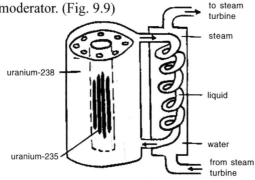

Fig. 9.9 Structure of breeder reactor

Which was the first nuclear submarine?

The first nuclear submarine was Nautilus, which was used by the U.S. Navy in 1958. It was launched on 21st January, 1954. The most modern nuclear submarine 'Ohio' of USA is in service since 1981.

What are cosmic rays?

Cosmic rays are streams of tiny particles that constantly enter the earth's atmosphere from outer space. Some of them originate in the sun but most of them come from beyond solar system. They are mainly protons and probably produced by storms on the sun and other stars.

What is a nuclear reactor?

A nuclear reactor is a device in which energy is produced with the help of nuclear fission reactions. Basically a nuclear reactor consists of fissionable material like uranium rods, control rods of cadmium, graphite rods as moderator, and heavy water as coolant. The heat produced by the nuclear fission converts water into steam. This steam drives a turbine which in turn drives a generator and electricity is produced. (Figure 9.10)

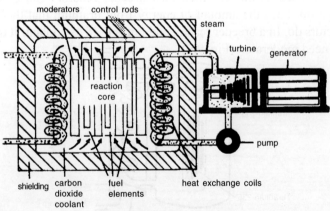

Fig. 9.10 Nuclear reactor

Which scientist built the first nuclear reactor?
Enrico Fermi built the first nuclear reactor in a Squash court at the University of Chicago, USA in 1942.

What is a Van de Graaff Generator?
It is an electrostatic generator used for accelerating charged particles. It consists of a tall metal dome at the bottom of which a device sprays electric charge on a rubber belt. The belt carries the charge to the top of the dome and transfers it to the dome's surface. Charges of several million volts can be built up in this way. (Fig. 9.11)

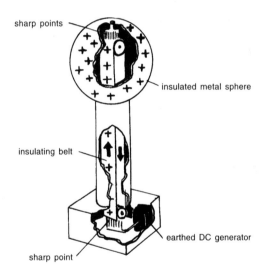

Fig. 9.11 Van de Graaff generator

How much more powerful is hydrogen bomb than the atom bomb?
Hydrogen bomb is 700 times more powerful than the atom bomb.

Which elements are used in hydrogen bomb?
Tritium and deuterium.

●●●

10. SPACE EXPLORATION

What is space?

Space is defined as the vast and empty region with no known boundaries and that exists beyond the earth's atmosphere. It apparently extends in all directions up to infinity and contains solar system, all the galaxies and the regions between the galaxies.

How does a rocket work?

A rocket works on the principle of action and reaction. According to Newton's third law of motion, for every action there is an equal and opposite reaction. The basic rocket engine contains a combustion chamber and exhaust nozzle. Modern rockets carry their own oxygen supply to burn their fuel and do not require any surrounding atmosphere. The burning of fuel escaping from the nozzle creates thrust by which a rocket moves. (Fig. 10.1)

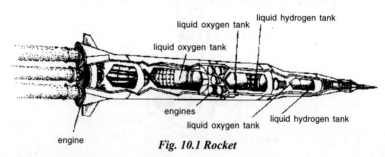

Fig. 10.1 Rocket

What are different rocket fuels?

Rockets burn different propellants, such as liquid hydrogen, liquid oxygen, nitrogen tetroxide, kerosene, solid fuels, etc. The rockets burn most of the fuel during the first few minutes.

What are multistage rockets?

Multistage rockets consist of two or more sections called stages. Each stage has its own rocket-engine and propellant supply. A multistage rocket can reach higher speeds than a single-stage rocket because it lightens its weight by dropping stages as it uses up propellants. For space exploration, multistage rockets have to be used which consist of a number of rockets joined together.

What are the different kinds of rockets?

There are four basic kinds of rockets; solid propellant rockets, liquid propellant rockets, electric rockets and nuclear rockets.

What is the difference between a rocket and a jet plane?

A rocket carries its own oxygen supply for burning the fuel whereas a jet engine takes air from the front of the engine.

What is a guided missile?

Any rocket whose flight-path can be altered in flight by guidance is known as guided missile.

What are the different types of missiles?

Missiles are rocket-propelled weapons. Modern missiles are often classified as surface-to-surface missile (SSM), air to air missile (AAM), surface to air missile (SAM) or air to surface missile (ASM). A cruise missile is in effect a pilotless computer-guided aircraft.

Which was the world's first ballstic missile?

The ancestors of modern missiles were the German V-1 and V-2, used to bombard London in World War II.

When did the real space age begin?

The real space age began on October 4, 1957, when world's first artificial satellite, Sputnik-1, was launched by Russia.

Who was the first man to enter into space?

Yuri Gagarin of Russia was the first man to enter into space on April 12, 1961, in the Russian spaceship, Vostok-1.

Who was the first woman to fly in space?

Russian lady Valentina Tereshkova was the first woman who went into space in June 1963. She made a three-day flight in the Vostok-6.

When was a living being sent into space?

The first living creature to go into space was a dog named Laika, carried in Russian satellite Sputnik-2 in 1957. (Fig. 10.2)

Fig. 10.2 Laika

What is an artificial satellite?

Artificial satellite is a man-made satellite which is launched into space with the help of rockets. They revolve around the earth in different orbits. More than 1500 artificial satellites are orbiting the earth.

Who was the first American to go into space?

Alan Shepard was the first American astronaut to go into space on May 5, 1961, in Mercury-3. (Fig. 10.3)

What is an orbit?

An orbit is the path of an object in space around another. For example, the planets move in orbits around the sun.

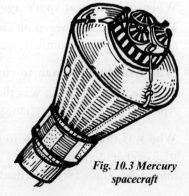

Fig. 10.3 Mercury spacecraft

What are the functions of the artificial satellites?

Satellites are doing many types of jobs for human beings. For example, communication satellites are to provide telephone and T.V. links throughout the world. Meteorological satellites are helping in weather forecasting. Navigation satellites are being used by ships and aeroplanes for different purposes. Scientific satellites are used to study earth resources, pollution, mineral deposits etc. They are also used for military purposes. (Fig. 10.4)

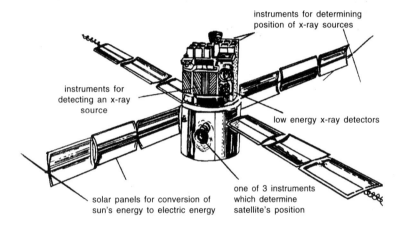

Fig. 10.4 An artificial satellite

Can an aeroplane fly in outer space?

No, because an aeroplane needs air to fly and there is no air in space. Beyond 160 kms. air becomes extremely thin, so an aeroplane cannot fly there.

What are the dangers in space?

The environment of outer space presents many hazards to astronauts. There is no air, great temperature extremes, radiations, such as X-rays and meteoroids are present there.

What is the heat-shield on a spacecraft?

When a returning spacecraft comes back into the earth's atmosphere it gets extremely hot due to friction. To protect the astronauts, the front end of the capsule is covered with a heat-shield made of special plastic. The shield heats upto about 2700°C. Some of the plastic melts and burns off taking away the dangerous heat.

What is meant by docking of two spacecrafts?

Docking means linking of two spacecrafts, in the orbit. The spacecrafts use their small rocket motors to line up and then slowly move towards each other until they can lock. This is called docking.

Describe a space suit.

A space suit is the covering of an astronaut which provides all survival requirements. It protects him from dangerous radiations and changes in air pressure when he is outside the cabin of the spacecraft.

How do astronauts get rid of body wastes?

Liquid waste is pumped into space, where it becomes a gas. Solid waste is put into plastic bags with chemicals that kill the germs. The bags are thrown away when spacecraft returns to earth.

Who was the first man to walk on the moon?

Neil Armstrong of USA was the first man to set foot on the moon from Apollo-II on July 20, 1969, saying —"That's one small step for a man, one giant leap for mankind".

Who walked in space first?

Edward H. White floated free outside the space vehicle Gemini-IV for the first time for 21 minutes on June 3, 1965.

Who remained in space for the longest time?

Yuri Romanenko of Russia remained in space for 326 days and came back to earth on December 29, 1987.

What is space walk?

When astronauts or cosmonauts go outside their orbiting spacecraft, we say that they are taking a space walk. Actually, they are not really walking, but are drifting alongside the spacecraft.

What is countdown?

A countdown is a check-up time before a spacecraft is launched from earth. During this time, every inch of rocket and spaceship is tested to see that it is in perfect working order.

A countdown may take hours or even days. It will continue until all the dials flash a green light. A loudspeaker booms "10-9-8-7-6-5-4-3-2-1-ZERO". With a loud roar the rocket blasts off and the spaceship begins to rise.

Can two astronauts talk on the surface of moon?

Two astronauts can't talk on the surface of moon because there is no atmosphere. For talking they use small radio sets which are built into each spacesuit.

What do astronauts eat?

They eat freeze-dried foods to save space and to keep things fresh.

What is splash down?

Splash down is the moment a capsule lands in the earth's water. As soon as splash down takes place, ships and helicopters rush to the floating capsule. The astronauts are lifted into a helicopter and are taken to nearby ship.

What is the difference between a cosmonaut and an astronaut?

Cosmonaut comes from Greek word meaning "Sailor of the universe". A cosmonaut is a Russian space traveller. An American space traveller is called an astronaut. The word means "Sailor among the stars".

What are space probes?

Space probes are unmanned space vehicles which have been sent to study Mars, Venus, Mercury and other planets.

What is a space station?

It is a special kind of satellite that can encircle the earth a few hundred miles up. Space stations are used for carrying out astronomical observations and surveys of earth, as well as biological studies and processing of materials in weightlessness. (Fig. 10.5)

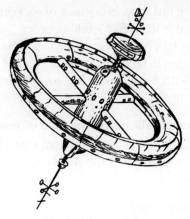

Fig. 10.5 Space station

What was sky-lab?

Sky-lab was a large and comfortable US Scientific space station in which astronauts set up duration records for living and working in space. It was launched in May 1973 to study the possibility of establishing space stations. It fell back on earth near Australia on 11 July, 1979. (Fig. 10.6)

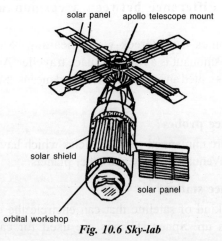

Fig. 10.6 Sky-lab

What is the Space Telescope?

Space telescope was launched in 1989 by Space Shuttle. It is ten times more powerful than any telescope. It is called Hubble Telescope.

What is a space shuttle?

A space shuttle is a reusable spacecraft which can be used like a plane for space missions. (Fig. 10.7)

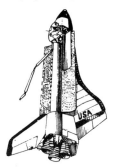

Fig. 10.7 Space shuttle

What is a space colony?

A space colony is a kind of island in space where thousands of people can live and work. No space colony exists yet, but the USA and Russia are planning to make such colonies. (See fig. 10.8)

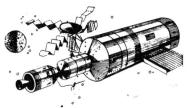

Fig. 10.8 Space colony

Who was the first man killed in space research?

Vladimir Komarov was the first person to be killed in space research, when his ship Soyuz-I crash-landed on earth on 24 April, 1967.

●●●

11. TIME

What were the earlier watches and clocks?
Sundial, sand clock, water clock and candle clocks were the earlier devices of time measurement. (Fig. 11.1)

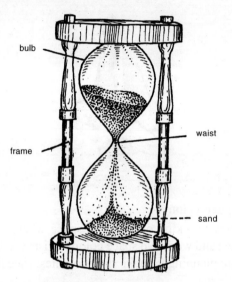

Fig. 11.1 Sand clock

What types of watches and clocks are used today?
Mechanical and electrical clocks. Mechanical clocks or watches are mainly spring-driven while electrical clocks are battery-powered.

How does a quartz clock work?
In a quartz clock a quartz crystal is used. It is a piezo-electric crystal. In a clock, the crystal is stimulated electrically to make it vibrate at its resonant frequency. It produces voltage at that frequency. It is used to drive a motor or a liquid crystal display which indicate time. It has great precision, with a short-term accuracy of about one-thousandth of a second per day.

How does a transistor clock work?

A transistor clock works by the amplifying action of a transistor. When the south pole goes into the detection coil, an induced current is generated in the coil that repels the south pole. This current is sent to the transistor. Part of the current in one direction is amplified to produce a magnetic field in the attraction coil. Thus the permanent magnet is attracted to the coil. This motion is repeated which measures the time. (Fig. 11.2)

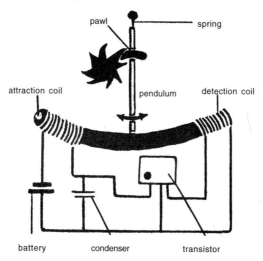

Fig. 11.2 Mechanism of a transistor clock

What is a chronometer?

It is a portable instrument used by navigators, surveyors and autoracers to measure time. John Harrison in England built the first accurate marine chronometer in 1761 capable of an accuracy of a half minute a year.

What is a sidereal year?

The time taken by earth to complete one revolution around the sun is called sidereal year. The sidereal year lasts 365 days, 6 hours, 9 minutes and 9.54 seconds.

What is a solar day?

The time taken by the earth to make one complete rotation on its axis is called the solar day. The 24-hour period between one midnight to the next is called the mean solar day.

What is standard time?

It is the time-system for various countries on an international basis. For this purpose the earth has been divided into 24 longitudinal zones each of 15 degrees of an arc corresponding to one hour. The zero zone is centered at Greenwich (London) which gives Greenwich Mean Time. The term Greenwich Mean Time (GMT) was replaced by Coordinated Universal Time (CUT). The zones to the east are numbered from 1 to 12 and west also are similarly numbered. Each zone is equal to 1 hr. (Fig. 11.3)

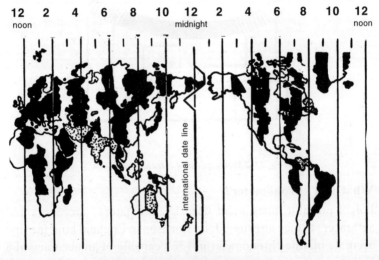

Fig. 11.3 International date line

What is International Date Line?

International Date Line (IDL) is a modification of the 180th meridian which marks the difference on time between East and West. The date is put forward a day when crossing the line going West and back a day when going East. The IDL was chosen at the International Conference in 1884.

What is universal time?

Since January 1972, a new standard of time called the Universal Time (UT) is being maintained in Paris by the Office of Weights and Measures. This is based on the average of the atomic clock readings from 18 timing centres around the world.

How is difference between the Universal Time (UT) and Coordinated Universal Time (UTC) is adjusted?

The difference between UT and UTC, based on earth's actual rotation, is adjusted by addition or subtraction of leap seconds on the last day of June and December.

What is atomic second?

The atomic second is defined as the time taken by the cesium electron to complete 9,192,631,770 spins. This has been taken as the unit of time.

How does an atomic clock work?

In an atomic clock, cesium metal is heated in a small oven. The cesium produces a beam which is directed through an electromagnetic field. The 5MHz output from a quartz clock is multiplied to give the 9192631770Hz that controls the electromagnetic field. Part of the 5MHz output is used to drive a clock display unit which indicates the time. It is so correct that it can lose or gain only one second in 3000 years.

What is sidereal time?

The time measured by the earth's revolution round the sun is called sidereal time. A sidereal day lasts 23 hours, 56 minutes and 5 seconds. Sidereal time is used in astronomy when determining the locations of celestial bodies.

What is time dilatation?

Virtual increase in time as per the theory of relativity is called time dilatation.

How many days are there in a leap year?
366 days.

Do all countries have the same time?
No, all countries have different times.

What is a digital watch?
It is a quartz watch which shows time in numbers. It is electrically operated.

12. CHEMISTRY

What is Chemistry?
Chemistry is the science concerned with the composition of matter (gas, liquid or solid) and of changes that take place in it under certain conditions. (Fig. 12.1)

What are the branches of Chemistry?
Physical, Inorganic, Organic and Biochemistry.

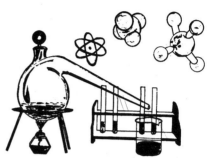

Fig. 12.1 Laboratory

Define (a) Deliquescence and (b) Efflorescence.
(a) It is a phenomenon of a substance absorbing so much moisture from the air that it ultimately dissolves in it to form a solution. Sodium hydroxide and calcium chloride are deliquescent.

(b) The loss of water from hydrated crystals, when they are exposed to air resulting in a dry powdery surface, is known as efflorescence.

What is a catalyst?
A catalyst is a foreign substance in a chemical reaction which alters the rate of reaction without undergoing any change in itself. In practice, most catalysts are used to speed up reactions.

What is a colloid?
A colloid is a mixture in which one substance in a finely divided state is dispersed in another substance. Blood, milk, ink, fog, etc. are examples of colloids.

What is the chemical name of marsh gas?
Methane gas.

What is the main use of Kipp's apparatus?

It is an apparatus mainly used for making and supply of H_2S gas. (Fig. 12.2)

What is sublimation?

It is defined as the change of a substance from solid to the vapour state without passing through the liquid phase. Dry ice or solid carbon dioxide, iodine, camphor and ammonium chloride are sublime substances.

What is the common name of brimstone?

Sulphur.

What is woods metal?

Woods metal is an alloy containing bismuth, lead, tin and cadmium. It has the extremely low melting point of about 70°C which is less than the boiling point of water.

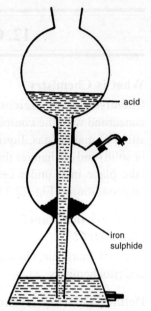

Fig. 12.2 Kipp's apparatus

What is the composition of air?

Air contains 78% nitrogen, 21% oxygen, 0.9% argon and 0.03% carbon dioxide. The remaining gases are helium, krypton, neon, ozone, water vapour, dust particles, methane, carbon monoxide, hydrogen and nitrous oxide.

What is TNT?

TNT is the short name for trinitrotoluene. It is the most commonly used explosive.

What are non-ferrous metals?

The term 'non-ferrous' is used to describe all metals other than iron. The most important non-ferrous metals are aluminium, copper, zinc, lead, tin and nickel.

What is the name of the furnace used for converting iron ore into iron metal?

Blast furnace. (Fig. 12.3)

What is a varnish?

A varnish is basically a paint without pigment. A varnish is a solution of resin dissolved in oil, which when applied to some wooden or metallic surface dries and becomes hard glossy and transparent coating.

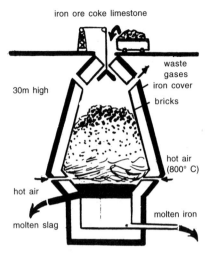

Fig. 12.3 Blast furnace

What is paint?

A pigment, either in dry form or diluted with oil, water or some other thinner for use in fine arts or as a protective covering for many surfaces.

What was the name of the first synthetic dye?

Mauveine.

What is the chemical name of gypsum?

This is calcium sulphate dehydrate. It ranks 2 in hardness on the Mohs' scale and is used for making casts and moulds and for blackboard chalk.

Which metals are used for making brass?

Brass is an alloy of copper and zinc with not more than 5% or 6% of other metals.

What are the six rare or noble gases?

Helium, neon, argon, krypton, xenon and radon. Noble gases were formerly known as inert gases. Radon is radioactive.

How many molecules are there in one cubic centimetre volume of air?

Under normal conditions of temperature and pressure 22.4 litres of air contain 6.023×10^{23} number of molecules. Therefore, one cubic centimeter of air contains about 2.5×10^{19} molecules.

What is water cycle?

It is defined as the journey of water from the oceans to the land and from the land back to the oceans. The heat of the sun evaporates water from oceans which takes the shape of clouds and again falls to the land in the form of rains and snow. From land it again goes to oceans, thus completing the cycle (Fig. 12.4). This cycle takes about 1000 years.

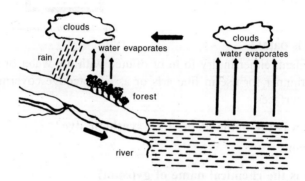

Fig. 12.4 Water cycle

Which metal does exist as liquid?

Mercury is the only shiny metal which exists as liquid. It freezes at –38.87°C and boils at 356.58°C. It is used in thermometers and barometers. It is also used in the preparation of many organic compounds.

What is stainless steel?

A broad term commonly used for an entire group of iron-based alloys that are remarkably resistant to rusting and corrosion. Stainless steel usually contains 70-90% iron, 12-20% chromium and 0.1% carbon.

What is dry ice?

Dry ice is the popular name of solid carbon dioxide. This is a dense snowlike substance. This is extremely cold having a temperature of about –78.5°C. When it melts it directly changes into gas without becoming liquid. It is used for packing ice-cream and meat, to prevent them from spoiling.

What is the full form of D.D.T.?

D.D.T. is the abbreviation of Dichloro-Diphenyl Trichloroethane. It is used to kill insects.

Which substance is used to preserve biological specimens?

Formaldehyde or formalin is used as preservative for biological specimens.

Who was Alfred Bernhard Nobel?

He was a Swedish chemist who invented dynamite and founded the Nobel Prizes from the money he earned from dynamite. These annual international prizes are awarded for outstanding work in physics, chemistry, physiology or medicine, literature, peace and economics. These prizes were first awarded in 1901 except for the prize in economics, which was instituted in 1969.

Why are metallic surfaces very shiny?

Metals contain free electrons. They make metals good conductors of heat and electricity. Free electrons are also responsible for the shiny appearance of most of the metals. They prevent the light from entering deep into the metal. The light gets reflected and, as a result, surface appears shiny.

Name the various allotropes of carbon, phosphorus and oxygen.

Carbon   – graphite and diamond.
Phosphorus – yellow phosphorus and red phosphorus.
Oxygen   – ozone

What is the mass of one litre of air?

A litre of air weighs about 1.3 gms.

What is another name for mercury?
Quicksilver. It is called so because of its silvery colour and fluidity.

What for is Bessemer process used?
It is a process used for converting pig iron from a blast furnace into steel.

What is bleaching powder?
It is a chlorine compound used for bleaching paper, pulps and fabrics and for sterilizing water.

What is petroleum and how was it formed?
Petroleum is a naturally occurring thick greenish-black liquid and is a mixture of complex hydrocarbons. Petroleum was formed over hundreds of millions of years as plants and animals died and decayed and were subjected to heat and pressure, within the earth's crust. A cutway view of an oil deposit is shown in figure 12.5.

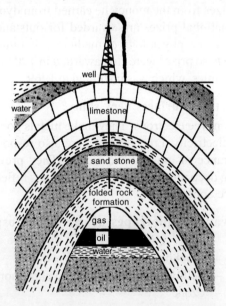

Fig. 12.5 This is a cutway view of an oil deposit in the earth

What is Canada balsam?

It is yellow tinted resin mixed with a volatile oil. It is obtained from the balsam fir tree. It is used for mounting specimens in optical microscopy.

By which thing is purity of gold measured?

Carat measures the purity of gold. Pure gold is described as 24 carat gold. The 14 carat gold contains 14th part in 24 of gold, the remainder usually being copper.

How is crude oil refined?

Crude oil is refined in a refinery by a fractionating column. The crude oil is heated until it evaporates. The vapour rises and the lightest parts reach the top and are piped out. These are fuel gas and petrol. However, fractions such as paraffin, heating oil and lubricating oil come out lower down the column. (Fig. 12.6)

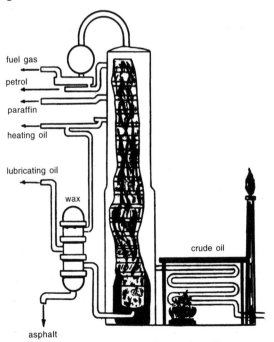

Fig. 12.6 Fractioning column of a refinery

What is LPG?

Liquefied Petroleum Gas used for cooking purposes.

How do we define boiling point?

The temperature at which a liquid begins to boil under normal atmospheric pressure. Boiling at this temperature starts because the vapour pressure above the surface of the liquid equals the atmospheric pressure.

Solid, liquid and gas are the three states of matter. Which is the fourth state of matter?

Fourth state of matter is called plasma state. In this state of matter, atoms and molecules exist as positive and negative ions in equal number. It is formed only at extremely high temperature. The shining material in the fluorescent tube light is an example of plasma state. (Fig. 12.7)

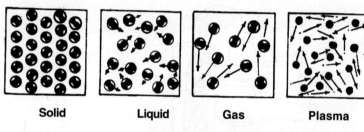

Fig. 12.7 Four states of matter

How is aqua regia made?

It is made by mixing one part of concentrated nitric acid and three to four parts of concentrated hydrochloric acid. It is a powerful solvent.

What is an alkaloid?

A group of nitrogenous organic compounds derived from living plants and having a powerful effect on human body. Important alkaloids are nicotine found in tobacco, opium and morphine of poppy plants and quinine in cincona.

What is Brownian movement?

It is the continuous random movement of small solid particles when suspended in a fluid medium. It was named after its inventor Robert Brown in 1827. This effect is also visible in particles of smoke suspended in a still gas. The suspended particles move because they are jostled by the constantly moving invisible atoms or molecules in the air or in the fluid. (Fig. 12.8)

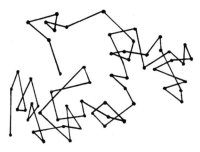

Fig. 12.8 Brownian motion

What are carcinogens?

Any agent that produces cancer is called carcinogen, e.g., tobacco smoke, certain chemicals and ionizing radiation, such as X-rays, ultraviolet rays and radioactive rays.

What are noble metals?

The group of metals like gold, silver, platinum, iridium, rhodium, osmium, ruthenium and palladium are called noble metals. They are called so because they are unreactive, corrosion resistant and expensive.

What are the important semiconductor materials?

Germanium silicon, selenium and lead telluride are important semiconducting materials.

Which process is used for copper extraction?

Bessemer Process.

What is the process of annealing?

It is a slow heating and cooling treatment applied to a metal to soften it, relieve internal stresses and instabilities and make it easier to work on machine.

What are the three main constituents of glass?

Three main constituents of glass are silica, sodium carbonate and calcium carbonate. When they are ground together and melted in a furnace, they produce glass on solidification. (Fig. 12.9)

sand 70% + soda 15% + limestone 10% + sundry substances 5%

Fig. 12.9 Main constituents of glass

Who was awarded Nobel prize for his contribution in chemistry in 1974?

Flory Paul John, an American Polymer chemist, was awarded Nobel Prize for his contribution to the science of polymer materials.

What is 'absolute zero' and its value?

Absolute zero is the lowest possible temperature. It is equal to −273.15°C (− 459.67°F). At this temperature the molecules of the substance would be completely at rest and the substance would possess no heat energy.

What is pasteurization?

Pasteurization is the process of preserving foods by heating them to kill the micro-organisms. It is commonly used for milk but may be used for cheese, butter, beer, etc. Milk is boiled at 72°C to 85°C for 16 seconds and then cooled quickly to 10°C. This kills most of the bacteria in the milk and allows it to stay fresh for several days.

What are the main methods used for leather tanning?
Vegetable tanning, chrome tanning and East India tanning are the three main methods of leather tanning.

What are acetals?
Acetals are the organic compounds formed by adding alcohol molecules to aldehyde molecule.

What is osmosis?
Osmosis is the process in which a solvent moves from a dilute solution to a more concentrated one across a semi-permeable membrane which allows the solvent molecules to pass selectively.

What is laughing gas?
Nitrous oxide is called laughing gas. It is used in anaesthesia.

Which metal can be extracted from red crystals of cinnabar?
Mercury.

What is the mineral from which common table salt comes?
Halite.

How is electroplating done?
The coating of a metal on another metal by means of electrolysis is called electroplating. The metal which is to be coated is made anode and the object on which it is to be coated is made cathode. A suitable solution of the electrolyte is put in the tank. Anode and cathode are connected with a battery. (Fig. 12.10)

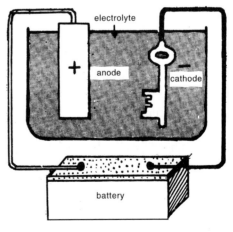

Fig. 12.10 Electroplating

Which mineral has high heat resistance?
Asbestos.

What is an alloy?
An alloy is a mixture of two or more substances at least one of which is a metal. For example, steel is an alloy of iron and carbon. Alloys are tougher and mechanically superior to the constituent metals.

What are natural fibres?
Wool, cotton and silk are natural fibres.

What are amino acids?
A group of organic acids having both carboxyl and amino group is called amino acid. There are 24 amino acids commonly occuring in proteins. Eight of these are obtained from the diet and cannot be synthesized by humans.

What is carbon cycle?
The carbon cycle is the process through which carbon is being continuously removed from nature and used and replaced by living beings. This keeps the percentage of carbon fixed in nature. Today carbon cycle is in danger of being disrupted by the increased consumption and burning of fossil fuels and destruction of forests. (Fig. 12.11)

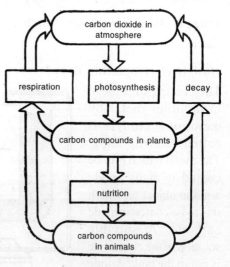

Fig. 12.11 Carbon cycle

What is an adhesive?

It is a substance used for joining two surfaces together. They are usually resins which are mixed with a curing agent to make them set. Araldite is a well known adhesive.

What type of forces exist in adhesion?

Vander Waals' forces. These forces are much weaker than chemical bonds.

Which class of adhesives set hard on cooling?

Thermoplastic.

What is ozone gas?

Ozone gas is a dark blue gas with pungent odour. It is an allotrope of oxygen with the chemical formula O_3. The upper atmosphere contains a layer of ozone which protects us from the ultraviolet rays of sun.

How is air liquefied?

Air is liquefied in machines. The temperature at which it boils is 198°C.

What are the two main gases of producer gas?

Carbon monoxide and nitrogen are the two main gases of producer gas.

What are the main gases in water gas?

Hydrogen and carbon monoxide are the main gases in water gas.

What are cohesive and adhesive forces?

Force of attraction between the same type of molecules is called cohesive force, while the force of attraction between molecules of different materials is called force of adhesion.

What is teflon?

This is a waxy material, used as a covering to prevent sticking on cooking utensils and industrial appliances. Its chemical name is poly tetra fluoroethylene.

What is nitrogen cycle?

The nitrogen cycle is defined as the continuous circulation of atmospheric nitrogen among the soil, water, air and living organisms and then back to atmosphere. (Fig. 12.12)

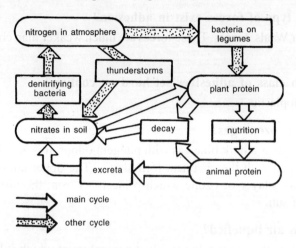

Fig. 12.12 Nitrogen cycle

Which material is used for making paper?
Wood and bamboo.

What is the full form of PVC?
PVC stands for Poly Vinyl Chloride.

Which plastic material is used for making gears?
Usually nylon is used for making gears.

What is a mineral?
The chemicals present in the crystaline form in many rocks are called minerals. In more general usage, a mineral is any substance economically viable for mining.

What is a metal ore?
Ore is a mineral from which the metal can be extracted profitably.

What makes metals a good conductor of heat and electricity?

The free electrons present in the metals make them good conductors of heat and electricity.

Why have metals high density?

The mass per unit volume is known as the density of a substance. In metals atoms are closely packed, so they have high density.

What is malleability?

Malleability is the property of a metal to be hammered or rolled into sheets. Gold is one of the most malleable metals. Other malleable metals are silver, copper, aluminium, tin, etc.

What is an electrostatic precipitator?

It is a device used in industry to clear dust and other particles from air. It uses electrostatic attraction to attract the particles to an electrode.

Define ductility.

The property of a metal that allows it to be drawn out into wire. Gold, silver, copper, etc. are ductile metals.

●●●

13. UNIVERSE

How and when did the universe originate?

The most widely accepted theory of how the Universe came into being is the 'Big-bang' theory. According to this theory the universe originated as a singularity, a highly dense entity where all the energy and matter of the present day universe was concentrated. Some 20 billion years ago, a tremendous explosion (big bang) occurred and matter was scattered in all directions. The cooling down of these scattered parts gave birth to galaxies and other constituents of universe.

Which element was the main component of the material given off by the big bang?

Hydrogen.

What are the constituents of universe?

The universe consists of the sun, earth, comets, galaxies and other celestial bodies of the present day.

How big is the universe?

The observable universe has a diameter of 25 billion light years.

What is a galaxy?

A galaxy is a cluster of stars, gas and dust held together by gravity. There are millions of galaxies in the universe. They are classified according to their shape and appearance. They can be spiral or elliptical. The diameters of galaxies range

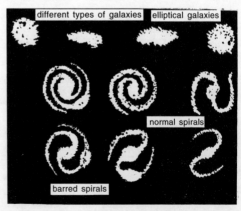

Fig. 13.1 Galaxies

from only a few thousand light years to 500,000 light years. Our solar system belongs to milky way. (Fig. 13.1)

How do we measure the distance of stars?

By the method of triangulation.

What is a Nebula?

A Nebula is a large cloud of gases and dust in outer space. Crab Nebula, Great Nebula and Horsehead Nebula are quite famous.

What is Crab Nebula?

It is a huge cloud of glowing gases that lies in the constellation Taurus. It resulted due to the violent explosion of a star called Supernova. It is about 6,500 light years from earth. The name comes from its crab-like shape.

What are nova and supernova?

A nova is a star that suddenly increases its brightness due to an internal explosion which helps it to throw off a small amount of star matter. It suddenly erupts in brightness by 10,000 times or more. A supernova is a much rarer one. It is thousand times brighter than a nova. In an explosion it throws off about one-tenth matter. Hence, its brightness increases by more than billion times. Very approximately, it is thought that a supernova explodes in a large galaxy about once every 100 years.

What is a quasar?

Quasar are quasi-stellar radio sources. They were first identified in 1963. Some astronomers think that quasars are very young galaxies. Quasars appear starlike, but each emits more energy than 100 giant galaxies.

What is a pulsar?

The term 'pulsar' stands for pulsating radio star. It is a rapidly spinning star that gives off rythmic pulses of radiation. The first pulsar was discovered in 1967 by British astronomers. Today more than 500 radio pulsars are known to exist in our Galaxy.

Which star becomes a neutron star?
Stars whose mass is between 1.2 times and less than 2 times the mass of the sun convert into neutron stars.

Which stars die as white dwarfs?
Stars lighter than 1.2 solar mass tend to die as white dwarfs. Most have surface temperatures of 8000°C or more and are hotter than the sun.

What is black hole?
A region in space whose gravity is so intense that nothing, not even light, can escape from it. A black hole is formed from the catastrophic collapse of a very massive star. (Fig. 13.2)

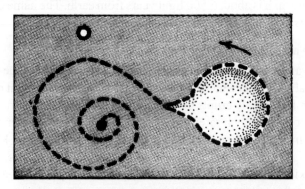

Fig. 13.2 Black hole

What is a binary star?
Binary star is a pair of stars moving in orbit around their common centre of mass. Binary stars appear single to the naked eye but in fact they are found to be double and can be distinguished by a telescope.

What is a red giant?
When hydrogen in a star is depleted, its outer region swells and reddens. This is called a red giant. Red giants have diameters between 10 and 100 times that of the sun.

What is a shift?

If a star or a galaxy is moving away from us, its light will be shifted towards the red-end of the spectrum. This is known as red shift. On the other hand, if a star is moving towards the observer its light will be shifted towards the blue-end of the spectrum. This is known as blue shift. This effect is known as Doppler effect.

Which is the brightest star in the night sky?

Sirus is the brightest star in the night sky. It is about 8.6 light years from the earth. It is also known as 'Dog Star'.

Which is the nearest star to us?

Proxima Centauri is the nearest star. It is about 4.25 light years away from the earth. It is a faint red dwarf, visible only with a telescope.

What is North Star?

The North Star or Polaris is a bright star. It always appears in the north because it is almost exactly above the North Pole. It is easily located because Dubha and Merak, two stars in the Big Dipper, point to it. (Fig. 13.3)

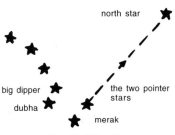

Fig. 13.3 North star

What is our solar system?

The sun, the planets and their moons, the asteroids, the comets and the smaller lumps of rocks (meteors) are all included in the solar system.

How far is the sun from the centre of our galaxy?

About 30,000 light years.

What is a sunspot?

A sunspot is a dark area that appears on the sun's surface, actually an area of cooler gas. Sunspots are several thousand degrees cooler than the surrounding area.

Name the nine planets of the solar system?

Nine planets of the solar system are Mercury, Venus, Earth, Mars, Jupiter, Saturn, Uranus, Neptune and Pluto. (Fig. 13.4)

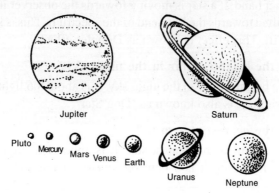

Fig. 13.4 Sizes of the planets

What is the temperature of the surface of the sun?

About 6000°C.

What is solar wind?

The solar wind is the stream of charged particles like protons, electrons, etc. from the sun's corona flowing outwards at speeds of between 300 kps and 1000 kps.

How many complete revolutions does the moon make around the earth in a calendar year?

Little more than thirteen.

Which planet has rings?

Saturn.

If you can jump one metre high on the earth, how high can you jump at the surface of moon?

At the surface of moon one can jump six times higher than earth because the gravity at moon is one-sixth that of the earth.

What are meteorites?

Meteorites are the interplanetary matter which enter the earth's atmosphere on being attracted by the earth's gravitational pull. They are primarily of three types: stony meteorites or aerolites, iron meteorites or sidrites, and intermediate called siderolites. Meteorites provide evidence for the nature of the solar system.

What are comets?

Comets are the shining heavenly bodies with tails. They belong to the solar system. Their tail is always opposite to the sun. They are composed of methane, ammonia, snow particles, etc. The brightest of them all is Halley's comet.

What are the phases of moon?

Changes in the moon's size are called the phases of moon. The size of the moon changes because it changes its position with respect to sun and hence its portion illuminated by sunlight also changes. These are called phases of moon. (Fig. 13.5)

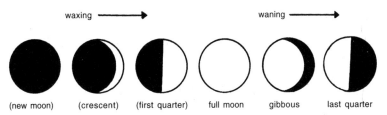

Fig. 13.5 Phases of the Moon

After how many years does the Halley's comet reappear?

The Halley's comet, named after the English astronomer Edmund Halley, reappears after every 76.3 years. It will next reappear in the year 2061.

What are asteroids?

Asteroids are the swarms of tiny planets revolving round the sun between the orbits of Jupiter and Mars.

Which is the smallest planet?
Pluto. It has a diameter of about 3,000 km.

Which is the hottest planet?
Venus.

Which planet is known as the red planet?
Mars.

Which planet has a great red spot?
Jupiter.

How old is the earth?
The earth is considered 4,600 million years old. Man appeared on it only one million years ago.

Name the various continents of the world?
There are seven continents, viz. Europe, Asia, Africa, North America, South America, Australia and Antarctica.

When did life originate on the earth?
The life started originating on earth some 570 million years ago. The first 345 million years saw the development of marine life. In the next 160 million years reptiles came into existence, and the subsequent 65 million years saw the development of mammals. The evolution of man is only one million years old.

What causes tides in seas?
Moon's force of gravity is mainly responsible for tides in seas. The sun's gravity also affects tides formation, but only to a limited extent. The difference between high and low tide is called "tidal range".

How many oceans are there?
There are five oceans namely Pacific, Atlantic, Indian, Arctic and Antarctic.

What is the inner structure of earth?

The upper layer of earth is called crust. Underneath the crust is a semifluid called mantle. Going further deep is the core made up of iron and nickel. (Fig. 13.6)

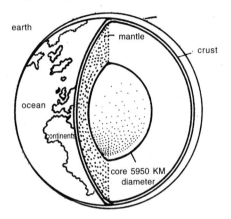

Fig. 13.6 Inner structure of earth

How much is the weight of the earth?

Earth's weight is about 6000 million million million tonnes. It has been calculated using Newton's law of gravitation.

How much water is there on earth?

It is estimated that hydrosphere contains about 1,460,000,000 cubic tons of water.

How many types of mountains are there?

Mountains are divided into four types, namely, fold mountains, clock mountains, volcanic mountains and residual mountains.

What are seismic belts?

Regions of earth which are prone to earthquakes are called seismic belts. Such regions are Western coast of America, Eastern coast of Asia, South Pacific Islands (Java, Sumatra, Indonesia).

How many types of earthquakes are there?
Earthquakes are of two types: volcanic and tectonic. Volcanic earthquakes are caused by the eruption of volcanoes, and tectonic earthquakes are caused by shifts in the rock structure of earth.

How much time does the earth take for completing one revolution round the sun?
The earth makes one complete revolution round the sun in 365.25 days.

What is the rotation period of earth on its axis?
Earth makes one complete rotation on its axis in 23 hours and 56 minutes. This is equal to the time of one day and one night.

How do the lengths of days and nights change during the years?
Lengths of the days and nights change due to the tilt of earth's axis. September 23 and March 21 have equal length of day and night. On June 21, day is longest and night is shortest, while on December 22, night is longest and day is shortest.

What does the sudden drop in atmospheric pressure indicate?
A drop in pressure often indicates the approach of stormy weather or rain. Areas of low pressure are called depressions.

What does a rise in atmospheric pressure indicate?
A rise in atmospheric pressure indicates a clear weather. The areas of relatively high pressure are called anti-cyclones.

What is a hot geyser?
It is a hot spring which throws up steam and hot water into the air. Geysers are common in Iceland and New Zealand. One of the best known geysers is 'Old Faithful' in Yellowstone National Park, USA.

What is geomorphology?
It is a branch of geological science that deals with nature and origin of surface landforms such as mountains, valleys, plains and plateaus.

What are doldrums?

Doldrums are the regions around the Equator with very low pressure and little wind. Sailing ships avoid the doldrums for fear of being becalmed.

What is light year?

The distance travelled by light in one year is called light year.

    1 Light year = 9.46×10^{12} km
    1 Par sec   = 3.26 Light years

What are naked eye stars?

The stars which are visible to the naked eye are known as naked eye stars. There are about 5000 naked eye stars. Our eye can see about 2500 stars at a time.

Which reaction is responsible for the energy generation in the sun?

Fusion reaction generates energy in the sun.

Who discovered uranus?

William Herschel.

●●●

14. PLANT KINGDOM

How many species of plants are there?
There are over 3,60,000 species of plants which have been studied by the scientists.

Into how many groups is the plant kingdom divided?
The plant kingdom is classified into four divisions depending upon their morphology and life cycle. These are: Thallophytes, Bryophytes, Pteridophytes, and Spermatrophytes.

Give examples of some plants belonging to these groups.
Algae and fungi belong to Thallophytes, liverworts and mosses belong to Bryophytes, selaginella, ferns belong to Pteridophytes and gymnosperm and angiosperm belong to Spermatrophytes.

When did plants appear on the earth?
A few small green plants like mosses and liverworts appeared on earth some 425 million years ago. About 400 million years ago some complicated plants came into existence. The first flowering plants developed about 136 million years ago, during 'Cretaceous period'.

How many species of flowering plants are there in the world?
About 2,50,000.

How are plants different from animals?
In general terms, a plant is distinguished from an animal by its ability to manufacture its own food by photosynthesis. But, animals cannot make their food. Most plants remain fixed at one place while animals are able to move from one place to another.

Which is the largest known flower in the world?
The largest known flower in the world is the Giant Rafflesia which grows in Malaysia and Indonesia. It measures up to 91 cm (3 ft)

across and 1.9 cm (³/4 inches) thick. The flower weighs between fifteen and twenty pounds.

What is the name given to the process of producing seeds by flowers?

This process is called pollination. When the pollen grains fall from one flower in the ovary of the other flower, pollination takes place which gives birth to seeds. (Fig. 14.1)

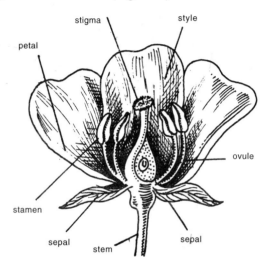

Fig. 14.1 Flower

Are there any plants that have neither flowers, fruits, seeds nor roots?

Plants can be divided into four major groups. One of these kinds is phyla which is made up of very simple plants called thallus plants or thallophytes. Thallophytes have no flowers, fruits, seeds, leaves, roots and stems. It includes algae and fungi.

Do mosses have flowers?

No. Mosses have leaves and stems but no flowers. They sometimes look as if they are blooming, but the 'flower' is really a little capsule full of spores at the end of a stalk or stem.

Why are leaves green?
Leaves of most plants contain a substance called chlorophyll. This substance gives the green colouration to plants.

Who first identified plant cells?
The English Chemist Robert Hooke first identified plant cells in 1665 through his microscope.

Which are the poisonous plants?
There are about 7000 species of poisonous plants. Some known poisonous plants are rhubarb, laurel, rosary pea, oleander, ivy, etc.

What is photosynthesis?
Photosynthesis is the process by which green plants make their food. Photosynthesis takes place in the presence of sunlight and chlorophyll that is present in green plants. In this process carbon dioxide of atmosphere is converted into starch and water. (Fig. 14.2)

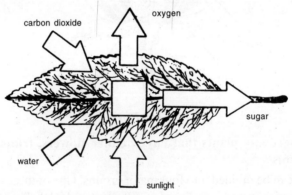

Fig. 14.2 Photosynthesis

Do plants sleep?
Yes, the plants sleep at night.

Why do some flowers close up at night?
The petals of flowers have a kind of automatic hinge arrangement. Some of these hinges are kept open by water pressure which the

sunlight helps it to soak up from inside the plant. When the light grows dim at night, the hinges lose water and grow limp. So the flowers close up.

How is the age of a tree determined?

The age of a tree can be determined by cutting a slice from the trunk of the tree and counting the number of rings on it. Each ring corresponds to one year. The number of rings counted from the centre of the trunk indicates the age of a tree. (Fig. 14.3)

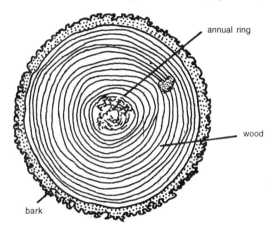

Fig. 14.3 Cross-section of the trunk

Which plants eat insects?

Insect eating plants are called insectivorous plants. A few such famous plants are: pitcher plant, sundew plant, venus fly trap, uticularia etc. A plant called Indian pipe, found in the hills of Shimla, is also insectivorous. (Fig. 14.4)

Fig. 14.4 Insectivorous plants

Which is the largest tree of the world?

The largest tree is the giant sequoia. Its name is General Sherman tree. It is located in Sequoia National Park in Central California. It measures 9.8 m in diameter. It is 85-metre tall and is estimated to be about 3,500 years old.

Which plant fibres are good for thread-making?

The important plants used for thread-making are flax, hemp and cotton. Linen is made from flax, ropes from hemp and cotton clothes from cotton fibres.

Why are the leaves of some plants sensitive to touch?

The leaves of some plants such as 'touch-me-not' are sensitive to touch. When we touch them the water from the thin-walled cells pass on to the stem. As a result, these cells shrink and lose their stiffness and curl up. (Fig. 14.5)

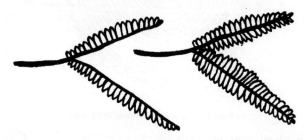

Fig. 14.5 Touch-me-not leaves

How do the leaves of a water lily float?

Water lilies often have leaves with many air spaces on the underside. The air trapped beneath the leaf makes it float on the water.

Why do some plants have thorns?

Thorns, prickles and sharp spines help to protect a plant from animals. They may also stop insects boring into the plants.

How many national parks and sanctuaries are there in India?

There are 62 national parks and 400 sanctuaries in India.

Which trees live the longest?

The world's oldest trees are the bristlecone pines. Some are believed to be over 450 years old.

Why do flowers have bright colours?

Flowers have bright colours to attract insects to carry out pollination.

Which is our national flower?

Lotus.

●●●

15. ANIMAL WORLD

What are the main divisions of animal kingdom?
The animal kingdom is divided into two main sub-kingdoms — Protozoa and Metazoa. Protozoa are unicellular and Metazoa are multicellular animals.

What is the simplest known form of animal life?
The amoeba. It is unicellular microscopic organism which lives in water and carries on all the life functions of higher animals.

When did fish evolve?
Fish first evolved about 530 million years ago.

How many species of fish are surviving today?
About 20,000.

Can a jellyfish be called a fish?
No. Jellyfishes are neither fish, nor are they made of gelatin. They belong to a large group of invertebrates that include sea anemones, corals, seafans and hydroids.

Does the shark sleep?
A shark never sleeps. (Fig. 15.1)

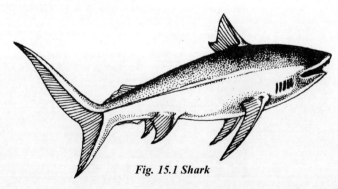

Fig. 15.1 Shark

Which mammals live in the sea?
Seals, dolphins and whales are all sea mammals.

How did electric fish get its name?
An electric fish is capable of giving a shock when touched, equivalent to 400 volts, sufficient to incapacitate any man. It is found in the region of the Amazon.

What is a starfish?
A starfish is not a fish at all, it is a beautiful sea animal with a body usually shaped like a five-pointed star, hence its name. The starfish has peculiar tube-like feet which it uses to open oysters. It has a remarkable property of regeneration.

What is the smallest known fish?
Dwarf Pygmy Goby (Pandaka Pygmea).

What is the peculiarity of an archerfish?
The archerfish shoots down its prey by squirting a jet of water at it. The water knocks the insect into water, the fish then eats it up. (Fig. 15.2)

Fig. 15.2 Archerfish

Do fishes produce sound?
It is generally believed that fishes do not produce sound, but they do produce sound and thus communicate with each other.

In what ways are whales like mammals?

They are warm-blooded. They have a heart in four parts. They breathe by lungs, give birth to youngs and feed them their milk. (Fig. 15.3). Whale is the largest living being.

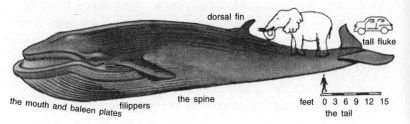

The size of blue whale compared with an elephant and a small car
Fig. 15.3 Blue Whale

Which creature has maximum number of ribs?
A snake.

Which is the largest kind of snake?
The anaconda.

Through what does a snake expel venom?
Fangs. (Fig. 15.4)

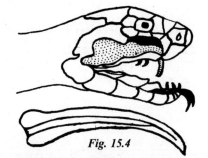

How can snakes eat animals bigger than themselves?
Snakes can swallow things that are much bigger than themselves because their jaw bones are loosely attached and the mouth can stretch to swallow a fat animal like a deer or pig. Its slim body can also stretch to make room for the animal.

Why don't snakes close their eyes when they are asleep?
Snakes don't have eye-lids.

What do snakes eat?
Some snakes eat other snakes; many eat birds, eggs or frogs; others eat insects, mice and crop-destroying animals.

Which sea animal squirts ink?
Squid squirts ink to protect itself.

How do snakes move along without feet?
They creep along by means of their ribs. Many of them can also climb up trees and steep places having rough surface.

What are carnivores?
The carnivores are the flesh-eating animals, like tiger, weasel, lynx, lion, wolf etc.

What are herbivorous animals?
Animals that eat grass, leaves, and different plants are said to be herbivorous or vegetarians. Some of them are: cow, rabbit, giraffe, goat, zebra, gorilla, etc.

What are omnivorous animals?
Animals which eat both meat and vegetables. They eat fruits, eggs, insects and grubs. Some such animals are badger, raccoon, rat, fox, ant, etc.

What is a saprophyte?
It is a plant, usually non-green, which absorbs food from decaying organic matter, e.g., yeasts, moulds, etc.

What is a parasite?
It is a plant or animal which absorbs or sucks its nutrition from other organism, and is physiologically dependent on them, e.g., worms, ticks, etc.

What are biorhythms?
Cycles that occur regularly in living beings are called biological rhythms, such as sleeping is a biorhythm which repeats every day. Heart beat is also a biorhythm.

What is bioluminescence?
Bioluminescence is the production of light by living things. The blinking of fireflies is an example of bioluminescence.

Which creature has the shortest life-span?

Mayfly, an insect, does not live more than 24 hours. Its sole function in life seems to be to perpetuate its species. For many mayflies die immediately after performing this function. (Fig. 15.5)

Fig. 15.5 Mayfly

Which creatures have the sharpest vision?

Birds. A vulture can spot a dead rabbit in the grass from a great height. A hawk can notice a tiny grasshopper from the height of 100 ft.

What is the science of studying birds called?

Ornithology.

Why do birds sing?

Birds sing to attract mates and to warn other birds away from their nesting area. Usually the male birds sing.

How do we define birds?

Birds are warm-blooded creatures of the class aves. They lay eggs and have feathers, wings and beaks. (Fig. 15.6)

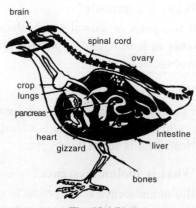

Fig. 15.6 Bird

Do birds have teeth?
No, they have beaks instead of teeth.

How many species of birds are there?
There are 8,733 known species of birds.

Which bird has been used as a messenger?
The pigeon.

Which bird lays only one egg in two years?
The albatross, one of the largest petrels. (Fig. 15.7)

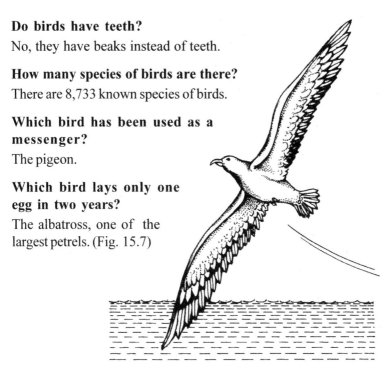

Fig. 15.7 The albatross

Which is the tallest bird?
The ostrich is the world's tallest bird. It can attain a height of 2.4 m. and weighs up to 150 kg. It cannot fly but can run very fast. (Fig. 15.8)

Which is the largest bird having the power of flight?
It is the wandering albatross; one albatross was found with a wing span of 11 feet, 4 inches.

Fig. 15.8 Ostrich

Which bird can fly backward?
The humming bird. It can fly backward, primary and side ways. It can also remain at one spot up to an hour. (Fig. 15.9). This is the smallest bird.

Fig. 15.9 Humming bird

Can any bird look direct into the sun?
Yes, eagle.

Is there any bird which never touches the ground?
Yes, the European swift never touches the ground. It rarely perches except at rest, even sleeping on the wings high in the air. It catches insects on the wing.

Is bat a bird?
Bat is not a bird but a mammal. It is warm-blooded. It has hair on its body and gives birth to its youngs and suckles them on milk produced by milk glands of the female bat. They do not lay eggs as other birds do.

What is a flying fox?
A flying fox, in fact, is a kind of bat whose head and face resemble with a fox. That is why it is called a flying fox.

Which animals make homes in (a) a burrow (b) a dray?
(a) Rabbit; (b) squirrel.

How can bats fly in dark?

The bat emits ultrasonic waves from its mouth while flying. The frequency of these waves is greater than the sound waves. After reflection from trees and buildings, etc., these waves reach the ears of the bats. By hearing they can judge the distance and direction of any obstacle in their way. (Fig. 15.10)

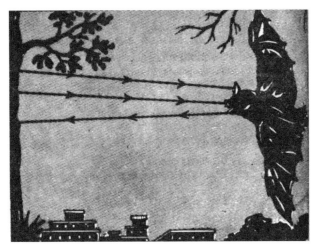

Fig. 15.10 Bat

What is produced by the mating of: (a) a mare and jackass (b) a mare and a stallion (c) two mules?

(a) A mule (b) a horse (c) nothing.

Where are polar bears found?

Polar bears are found in Arctic Ocean.

Can a fly hear?

The greater number of insects, including flies, cannot hear. Bees, ants and wasps are all deaf.

How is the silk produced?

Silk is produced by a kind of caterpillar called silk worm.

What is the name for (a) a rabbit's tail (b) a fox's tail?
(a) Scut; (b) brush.

True or false: (a) only a female bee stings (b) only a female wasp stings?
Both the statements are true.

How many wings do bees have?
Four.

How many pairs of legs do insects possess?
Three pairs of legs.

How does the movement of a cat's jaw differ from that of dog's?
Cat's jaw moves up and down, not sideways. A dog's jaw moves in either direction.

Where does the largest coral formation take place?
Great Barrier Reef, along the north-eastern coast of Australia.

How much is a snail's pace?
According to the National Geographic Society of USA, a snail travels 23 inches per hour.

Do all sheep in a flock habitually sleep at the same time?
No. It is a flock instinct that some sheep watch while others sleep.

Which is the largest living reptile?
The estuarine or salt water crocodile. It measures up to 14-18 ft. in length and weighs 408-520 kg.

To which creatures are the following noises attributed (a) baying (b) croaking (c) braying?
(a) Hound (b) frog (c) donkey.

Do some mammals lay eggs?
Yes. The duck-billed platypus and the echidna, both of which live in Australia, are the only egg-laying mammals.

Which animal can run the fastest?

The cheetah, a member of the cat family. Its initial burst of speed is amazing. Tested against a motorcar it was found that the animal could cover a distance of half a mile in 20 seconds, and a 100 yards in 4-1/2 seconds which is faster than the fastest greyhound. (Fig. 15.11)

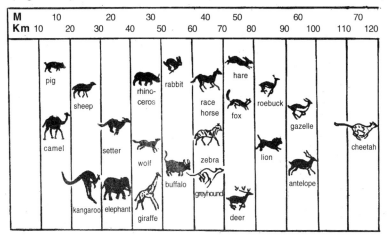

Fig. 15.11 Mammals

Which birds are the fastest fliers?

Swifts.

What are amphibians?

Amphibians are a group of vertebrate animals which are equally at home on land and in water. Their skin has no hair or scales, e.g., frogs, toads and salamanders.

How do insects without lungs breathe?

Oxygen from the air goes direct to each cell through a network of tubes and millions of branches which the zoologists call tracheal system.

What are tylons and ligers?

These are the hybrids between lions and tiger. The tylon has a tiger father and lioness mother; the liger has a lion father and tigress mother.

Which animal has the largest proboscis?
The elephant.

Which animal produces musk?
Musk deer produces musk. It is obtained from the musk pods contained in the sac under the skin of the male's abdomen. (Fig. 15.12)

Fig. 15.12 Musk deer

Is there any animal which does not have red blood?
Yes, some animals without backbone have blue blood due to hemocyanin. The blood of insects can be green, yellow or colourless.

What are rodents? Name some important rodents.
A rodent is a gnawing animal. They have long front teeth for gnawing. Some important rodents are squirrel, guinea pig, harvest mouse, porcupine, beaver, kangaroo, mouse, etc.

What are fossils?
Fossils are the hardened remains or imprints of plants and animals that lived millions of years ago. (Fig. 15.13)

Fig. 15.13 This is a fossil leaf which has been embadded in coal

What are scavengers? Name some of them.

Scavengers are those birds which feed on dead animals or the remains of the animals killed by others. Some important scavengers are: vulture, condor, lammergeier and carrion crow.

Does any animal have three hearts?

There is a peculiar animal called cuttlefish which has three separate hearts.

Which animal never drinks water in its entire life?

There is a kind of rat called kangaroo rat which never drinks water in its entire life. It is found in south-western deserts of the U.S.A. (Fig. 15.14)

Fig. 15.14 Kangaroo rat

Which are arboreal animals?

Animals which spend their lives high up in the tree-tops are called arboreal, such as birds, many type of monkeys, some lizards, etc.

Does an elephant drink water through its trunk?

No, it fills its trunk with water and then squirts it into its mouth.

Which animals hibernate during winter?

Hedgehog, marmot dormouse, bat, squirrel, adder, lizard, frog, tench, snail, etc.

Where are penguins found?
Penguins are found in the Antarctica. (Fig. 15.15)

What is a marsupial?
A marsupial is a primitive mammal which carries its young one in a pouch. After birth the young marsupials, which are not completely developed at the time of birth, continue to develop for some time in the mother's pouch. Pouched mammals include the kangaroos, wallabies, koalas, wombats, bandicoots and phalangers of Australia, and the opossums of America.

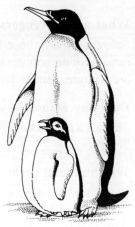

Fig. 15.15 Penguins

What are the peculiarities of kangaroo?
Kangaroos fall under the class of marsupials which means animals with pouches. It is about six feet tall and can cover a distance of 7 to 8 m. in one jump. The strangest fact about this animal is that when a baby kangaroo is born it looks like a naked mass of about 2.5 cm. in length and weighs only one gram. (Fig. 15.16)

Fig. 15.16 Kangaroo with baby

A group of lions is called a pride of lions. What would you call a lot of geese, partridges, monkeys, bees, porpoises, hounds, fish, sheep, starlings, cows and foxes?

1. A gaggle of geese.
2. A covey of partridges.
3. A troop of monkeys.
4. A swarm of bees.

5. A school of porpoises.
6. A pack of hounds.
7. A shoal of fish.
8. A flock of sheep.
9. A murmuration of starlings.
10. A herd of cows.
11. A skulk of foxes.

Which animal travels upside down?

Sloths, which live in the forests of Central and South America, hang down by his hooked claws while walking about the branches of the trees.

Are the fore legs of a giraffe longer than its hind legs?

No. It is just an optical illusion: the long neck, high shoulders, and the fact that the hind legs are slightly bent, all combine to give the impession that fore legs of these mammals are longer than they really are.

Are cat's whiskers useful to the animal or are they purely ornamental?

They are delicate sense organs, helping the animal to find its way about.

Which animals have only lower set of teeth?

Goat, cow, etc.

Which animal, when touched, drops its tail?

Lizard.

Which animal is dumb?

Giraffe. (Fig. 15.17)

Which animal is rated next to man in intelligence?

Chimpanzee.

Fig. 15.17 Giraffe

Which animal has the longest life span?
A tortoise may live as long as 500 years.

What is a ruminant?
A cud-chewing animal like cow, sheep, etc.

How does the cow's digestive system work? How many stomachs does a cow have?
Cows, sheep, goats, deer, camels, etc., belong to a group of animals known as the ruminants that eat grass and leaves which are hard to digest. These animals have a special stomach which has four chambers. The first chamber stores food. In the second, food is formed into balls or cuds and sent back to the mouth for thorough chewing when the animal is resting. Then it is sent to the third chamber which is filled with folds of tissues called the manyplies. The food filters past the folds and then enters the fourth chamber which is like the human stomach. (Fig. 15.18)

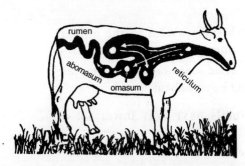

Fig. 15.18 Cow's digestive system

What is the difference between the horns of a cow and a deer?
The horns of a cow have the hardened outer layer of the skin and are retained throughout life. The horns of a deer are called antlers and are true bone. They are shed and regrown periodically, generally once a year.

Do cows have upper front teeth?
No.

What are the constituents of cow's milk?

Cow milk contains 87.2% water, 3.7% fat, 3.5% proteins, 4.9% sugar and many minerals and vitamins. Main minerals are calcium, phosphorus, iron, copper, manganese, sodium chloride, iodine, cobalt, etc. Main vitamins are A, B, C and D. The percentage of these constituents is different in the milk of various other animals.

Which fishes breath air?

Lung fishes are able to breath air and survive out of water.

How many teeth does a horse have?

A female horse has 36 teeth and a male horse 40.

How many toes does a horse have?

The horse, the donkey and the zebra have only one toe on each foot. The side toes are represented only by small splint bones.

How do (a) the horse and (b) the cow rise to their feet?

(a) The horse rises first on its fore legs;
(b) The cow rises first on its hind legs.

In which continents have the number of mammals and birds become extinct since 1600 A.D.?

A continental table showing a number of extinct mammals and birds since 1600 A.D. is given below:

Continents	Mammals	Birds
Africa	11	0
Asia	11	6
Australia	22	0
Europe	7	0
North America	22	8
South America	0	2
Total	**73**	**16**

Which animal has the largest teeth?

Elephant – tusk.

Which animal has the largest eye?
Giant squid – 15" dia.

How does a camel conserve water?
A camel conserves water in its hump.

What is the function of Queen ant?
To lay eggs.

How many bees live in a hive?
About 50,000 to 60,000 bee workers.

Which birds cannot fly?
Ostrich, emu, penguins cannot fly, but they are fast runners. Penguins are very fast swimmers.

●●●

16. HUMAN BODY

Which are the four primary divisions of the human body?

The head, neck, trunk and extremities.

Which elements constitute our body?

Our body is composed of oxygen, carbon, hydrogen, nitrogen, calcium, phosphorus, sodium, chlorine, sulphur, magnesium, iron and some other minerals in very small quantities.

What is meant by the organs of the body?

The organs of the body are the specialised body structures which perform special functions. Thus, the ear is the organ of hearing, the intestine is the organ of digestion and so on. The work carried out by a particular organ is called its function.

What is a vital organ?

Some of the organs of the body, such as the heart, brain, lungs, etc., are so closely associated with life of the body, that they are called 'vital' organs. If any of these organs be diseased or damaged, the health is adversely affected. The other parts of the body suffer in consequence, and if the trouble is not removed, death follows.

How do our eyes see objects?

Light rays coming from any object enter into our eye. A lens inside the eye produces an inverted image of the object on retina. This image is carried to the brain in the form of electrical impulses by optic nerve. The brain again inverts the image and it becomes erect. This is how we see objects. (Fig. 16.1)

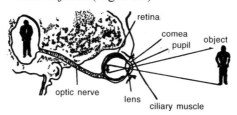

Fig. 16.1 The eye

Why do we blink?

We blink our eyes to keep them clean. As our eyelids close over our eyes, they wipe a layer of water over the surface. This clears away any dust on the eyeball. The transparent covering of the eye ball is called cornea.

What is the function of kidneys in our body?

Kidneys work like filters and remove the waste materials from the blood. They nearly clean 1700 litres of blood everyday. (Fig. 16.2)

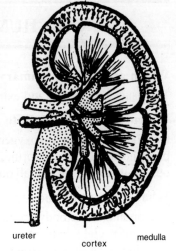

Fig. 16.2 Kidney

How does the tongue detect the taste?

Our tongue has the granular lumps on its upper side. These are called taste buds. These detect different tastes. The front portion has the taste buds which are sensitive to sweet and salty tastes. Back portion detects bitter taste while the buds on the edges are sensitive to sour taste. Taste buds are connected with brain by nerves which carry the information to the brain to detect a particular taste. (Fig. 16.3)

Fig. 16.3 Tongue

How many taste buds are there on the tongue?
We have about 1000 tiny taste buds on our tongues.

What is saliva used for?
Saliva is the fluid in our mouth and it has several uses. It helps us to swallow food by making our throat slippery. It also helps to dissolve food and also to digest food easily.

How do we hear?
Sound waves coming from some objects enter our ear and strike the ear drum. It starts vibrating. Three small bones transmit the vibrations to the inner ear. The vibrations make fluid in the cochlea move and electrical impulses are generated which go to the brain and we hear the sound. (Fig. 16.4)

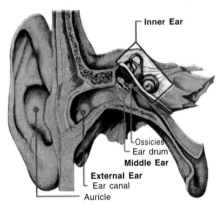

Fig. 16.4 Inner parts of the ear

What determines the sex of a child?
We know that ovum of the female and the sperm of the male contain sex chromosomes. Ovum has only one type of chromosomes called X-chromosomes but a sperm contains two types of chromosomes called X and Y chromosomes. They are named so because of the way they look through a microscope. If during the process of conception X chromosome from the sperm fertilizes the ovum, a girl is born. If however, a Y chromosome fertilizes the ovum, male child is born.

How do we breathe?

We inhale air through the nose or mouth. The air goes to the lungs through the wind pipe. The oxygen in the air is circulated throughout the body by the blood. It gets oxidised and converted into carbon dioxide which comes out when we breathe out. (Fig. 16.5)

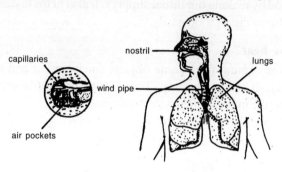

Fig. 16.5 Breathing system

What is the first thing a new born baby does?

A new born baby has to breathe for itself so it cries to start breathing.

What is the number of chromosomes in a nucleus?

There are 46 or 23 pairs of chromosomes in a nucleus.

How are twins born?

When the sperm of a man enters the ovum of a woman the fertilization takes place and the woman becomes pregnant. If by chance an ovum divides itself into two parts, the woman delivers two identical children either both boys or both girls. They are called identical or uniovular twins. In another possibility if two sperms from the male semen fertilize two ova of the same woman again two children are born. They are called unidentical or fraternal twins, because two separate ova have been fertilized. The two children may or may not be of the same sex. There are cases where women have delivered up to ten children at a time.

What are the three main types of twins?

(a) Identical or uniovular,
(b) Unidentical or fraternal, and (c) Conjoined or Siamese.

Where is the milk produced in female body?

Milk in females is produced in the mammary glands which are bag-shaped and located in the breast. (Fig. 16.6)

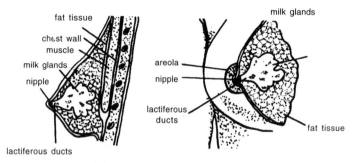

Fig. 16.6 Inner parts of the breast

At what age does puberty usually begin?
At the age of 12 to 13.

What is the full name of appendix?
Its full name is vermiform appendix. It is found on the lower right side of the abdomen where our small intestine joins the large intestine. (Fig. 16.7)

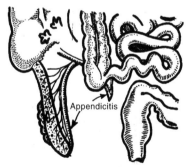

Fig. 16.7 Appendix

What is the thickness and area of human skin?
Thickness of human skin varies from 0.5 mm on the eyelids to 6 mm on the feet. The surface area of normal human skin is about 18 square feet.

What is the function of the pancreas?

Pancreas produces a strong digestive juice that breaks down food particles. It also produces insulin and glycogen. (Fig. 16.8). Insulin controls sugar level in the body.

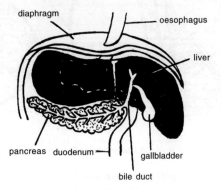

Fig. 16.8 Pancreas

What are freckles?

Freckles are little spots of colour in the skin. They show up when the skin is exposed to the sun because they are tiny patches of skin that tan more quickly than the surrounding skin.

What is the name of the oily secretion produced by glands in the skin?

Sebum (Fig. 16.9)

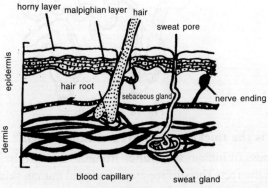

Fig. 16.9 Skin

Which skin pigment protects it from the harmful effects of sunlight?

Pigment melanin protects the skin from harmful effects of the sunlight.

What are the constituents of our blood?

Our blood is composed of four parts plasma, white and red blood cells and platelets. Platelets arrest bleeding by sticking to the edge of the wound. The circulation of blood through the body provides a mechanism for transporting substances.

What are the different blood groups?

Blood of all humans is mainly divided into four groups: A, B, AB and O. Group AB is called universal acceptor, while O is called universal donor. Group A can be given to persons having A and AB blood group. B can be given to B and AB. O can be given to anybody and AB group to persons of only AB group.

A thermometer is used to measure body temperature, a stethoscope for heart examination. What for is a sphygmomanometer used?

To measure blood pressure. (Fig. 16.10)

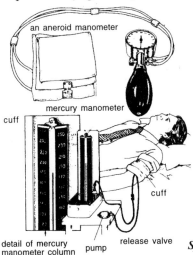

Fig. 16.10 Sphygmomanometer

123

How much blood is there in our body?
The amount of blood one has depends on how much one weighs. A fully grown person has about 5 litres of blood.

Why is blood red?
Blood is red because of the presence of a pigment in blood called haemoglobin.

Where do red blood cells originate?
In bone marrow.

What are vitamins?
Vitamins are group of organic substances which are essential for normal metabolism. Important vitamins are A, B, C, D, E and K. A, D, E and K are fat soluble while B & C are water soluble. Vitamins protect the body from diseases.

How many bones are there in the human body?
There are 206 bones in an adult body.

From which materials are human bones made?
Human bones are made up of mainly calcium and phosphorus.

Which is the longest bone in the human body?
The femur bone (thigh bone) is the longest bone in the human body. It constitutes usually appx. 28 percent of a person's stature.

Which material constitutes our nails?
Our nails are made up of keratin which is a kind of dead protein.

What is a hearing aid?
A hearing aid is an electronic device that improves the hearing ability of a hard of hearing person by amplifying sound. They are of two types: air-conduction aids and bone-conduction aids.

What is the audible frequency of human ear?
Below 20 Hz are called infrasonic and above 20,000 Hz are called ultrasonic.

Why do we sneeze?
Sneezing is a reflex action to expel air very quickly from the nose and mouth. We want to sneeze when our mucous membranes are irritated or swollen by cold.

Why do we get hiccups?
Hiccups are generally caused by not digesting food properly. Indigestion may irritate the diaphragm and the diaphragm then jerks making us take a sudden sharp breath and we hiccup.

Why do we cry when we peel onions?
Onion contains an oil called allyl. When we cut an onion, this oil escapes into the air and reaches our eyes and nose. This creates an irritation in the nerves of eyes and nose. These nerves send a message to the brain and as a result the brain responds with tears to soothe the irritation.

What do you understand by milk teeth and permanent teeth?
Humans have two sets of teeth. The first set of teeth which begins to appear at the age of about seven months is called milk teeth. These are 20 in number. The second set consisting of 32 permanent teeth begins replacing milk teeth at about six years of age.

Which is the hardest substance in the human body?
Tooth enamel. (Fig. 16.11)

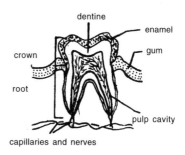

What is the normal breathing rate of a human being?
An adult normally breaths in and out about 20 times a minute. Children usually breathe faster.

Fig. 16.11 Structure of a tooth

Who was the first psychologist to determine a scale for measuring intelligence.
Alfred Binet, a French psychologist, who collaborated with Theodore Simon in devising the Binet-Simon tests, which are widely used to assess intelligence.

What is catabolism?
Catabolism is the metabolic breaking down of complex molecules into simpler ones in living organisms along with release of energy.

What is metabolism?
It is the sum of all the chemical and physical processes that occur within the body. Metabolism includes anabolism, which is the building up of complex organic substances from simple ones and catabolism, which is the breaking down of complex substances to simple ones with the liberation of energy.

What is atavism?
The resemblance of offspring to a remote ancestor.

What is the normal temperature of human body?
The average normal temperature of human body is 98.4° Fahrenheit (taken orally), equivalent to 37°Celsius.

Does the body temperature remain constant throughout the day?
The temperature normally fluctuates in the course of 24 hours from 97.3°F in the middle of the night to about 99.1°F in the middle of the afternoon. In addition, temperature differences occur between various areas of the body being lower in the skin.

Name the person who first conducted successful laboratory experiments on the conditioned reflexes.
Ivan Pavlov.

Which substance is stored in liver and muscles?
Glycogen. It acts as an energy reserve for the body.

What is Neurology?
It is the study of the nervous system, brain and its disorders.

What vitamin is available to the human body through sun bathing?
When sunlight containing ultraviolet rays falls on the skin, a substance (dehydrocholestrol) found in the body is converted into vitamin D and taken into the blood stream.

What is a reflex action?

It is an action that is not controlled by our will or by thinking. It is an involuntary response to a stimulation of the nerves as in shivering or sneezing. The common habit we have of 'winking' is an example of reflex action. (Fig. 16.12)

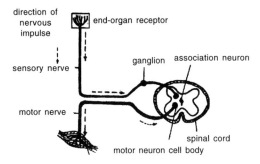

Fig. 16.12 Reflex action

How do we detect smell?

The smell is detected by the cells present in the olfactory nerves situated at the end of the nasal cavities. They are called receptors. When particles of the smelling material reach these receptors, they send electric impulses to the smelling centre of the brain and we detect the smell.

Why do we feel thirsty?

Our body contains salt and water in a fixed proportion. When the amount of water in the blood is reduced due to some reason, the proportion of these two materials changes. In such a situation the 'thirst centre' of the brain sends a signal to the throat by which it starts contracting. This makes it dry and we fell thirsty.

What are the tonsils?

Two flat glands of lymphatic tissue at the back of the throat. The tonsils are part of body's defence system, catching germs before they pass deeper into the body.

What is the average normal pulse of an adult male?

72 beats per minute.

What are ductless glands?

Ductless or endocrine glands are groups of cells which produce hormones for secretion directly into the bloodstream. These are: pituitary, thyroid, parathyroid, thymus, adrenal, panacreas, testes (in men) and ovaries (in women). They secrete about 25 hormones. (Fig. 16.13)

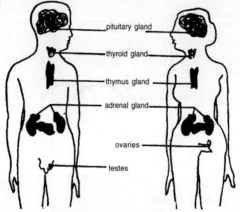

Fig. 16.13 Ductless glands

Does the pulse remain the same throughout life?

No. It slows down as age advances. At birth it is between 130 and 140 per minute, at 10-20 years it decreases to 80-85 and as adult it is 72 and after 60 it is in between 60 and 65.

Can a pulse be taken only at the wrist?

No. The beat can be felt at any large artery which is near the surface of the skin and rests directly on a bone.

What is the function of liver?

It is a very important organ of the body with many functions critical in regulating metabolic processes. They include the digestion of food, excretion, storage and conversion of food materials, regulation of blood composition and destruction of poisonous substances. (Fig. 16.14)

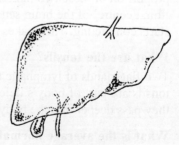

Fig. 16.14 Liver

Is the pulse, during low blood pressure, more or less rapid than normal?

More rapid, as the pulse increases in order to increase the output of blood.

What is filtrum?

It is the name given to the groove between your nose and your lips.

What is the common name for clavicle?

Collarbone. It forms a part of the shoulder gridle of the skeleton.

Which is the most complete food from the nutritional standpoint (a) leafy vegetable (b) root vegetable (c) legume?

Leafy vegetable.

What is a simple name for lactose?

Milk-sugar found only in mammalian milk and produced by the mammary glands.

What causes shock after an accident?

Stagnation of blood in the abdomen making brain and heart under supplied.

What are the main parts of our brain?

Our brain is divided into three main parts: Cerebrum, cerebellum and medulla oblongata. (Fig. 16.15)

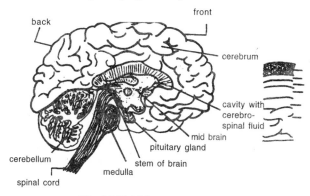

Fig. 16.15 Main parts of human brain

In which part of brain is language ability stored?
In the left hemisphere of brain.

In which area of brain are messages decoded?
Reticular formation.

How many nerve cells are there in human brain?
10^{10}.

At what age does brain attain maximum weight in (a) the male and (b) the female?
(a) About 20 and (b) about 17. The average male brain weighs 1.4 kg and the female 1.25 kg.

What do you understand by psychiatry and neurology?
Psychiatry is the study of mental diseases, their diagnosis, treatment, managements and prevention. Neurology is the subdiscipline of medicine that involves the study of brain, spinal cord and peripheral nerve, their diseases and conditions, and treatment of these conditions.

In psychiatry, what is the distinction between psychosis and neurosis?
Psychosis is a mental disease or any serious mental derangement; neurosis is a functional nervous disorder and is usually considered less serious of the two.

What is narcotic?
A drug that produces sleep or stupor, simultaneously relieving pain, such as opium and its derivatives; cocaine; barbiturates; and marijuana. Sedatives narcotics have an important place in medicine. But self-medication with any narcotic, for any purpose, is dangerous and damaging. Narcotics, such as morphine and heroin, may lead to addiction and may cause much physical and mental torture. The monetary pleasure and relief given by the drug can never compensate for the depression.

Who is the tallest man in the world?
Mohammad Alam Channa of Pakistan is the tallest man of the world. His height is 9 feet and 3 inches. He weighs 491 pounds.

Which parts of the body can be replaced with artificial parts.
More than 40 parts are made artificially and are replaced. These are heart valves, joints, arms, pacemaker etc.

Why does our hair go grey?
Our hair contains a pigment. When the pigment disappears, our hair goes grey.

Why do we need vitamins and minerals.
These are the protective foods and protect the body from diseases.

Why are vaccinations given to a person?
A person given a vaccine is made immune to the disease and cannot pass it on to anyone else.

How many diseases are controlled by vaccinations?
About 18 diseases.

What are the diseases controlled by vaccinations?
Mumps, chicken pox, hepatitis, rabies, cholera, typhus fever, yellow fever, plague, influenza, dihptheria, whooping cough, tetanus, small pox, polio, typhoid fever, measles, german measles etc.

17. HUMAN DISEASES

Name some important viral diseases.

Viruses are micro-organisms which range in size from about 0.01 to 0.3 micron (one millimeter of a metre). They spread many diseases, such as common cold, influenza, polio, small pox, chicken pox, mumps, measles, shingles, hepatitis, trachoma, rabies, AIDS, etc.

Which are infectious diseases?

Infectious diseases are those which are caused or transmitted by viruses, bacteria, fungi, protozoa, or worms. Similar to infectious diseases are contagious diseases which spread from one person to another by personal contract.

Which are bacterial diseases?

Bacteria are unicellular micro-organisms. They reproduce through the process of cell division i.e., one cell divides itself into two. They are mainly of four types—coccus (spherical egg-shaped), bacilus (rod-shaped), spirillum (spiral-shaped) and vibrio (comma-shaped). They spread typhoid, tetanus, T.B., cholera, diphtheria, dysentery, whooping cough, etc. (Fig. 17.1)

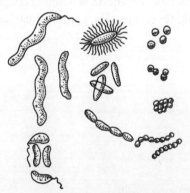

Fig. 17.1 Different types of bacteria

What is meant by the incubation period of a disease?

The period between the entry of germs into body and the appearance of actual symptoms.

Which are deficiency diseases?

Deficiency diseases are those which are caused due to the inadequate intake of nutrients, vitamins and minerals in our food.

What is the distinction between endemic and epidemic diseases?

Endemic is the term used to describe, for example, 'a disease that is indigenous to a certain area, owing to local conditions'. Epidemic disease is a violent outbreak of a disease affecting large number of people at one time and place, and is capable of spreading from one place to another.

What is Addison's disease?

It is a hormone deficiency disease. It causes weakness, digestive problems, heart problems and a brown colouring of skin. It is caused when adrenal glands do not produce enough cortical steroid hormones. It is treated by taking the hormone cortical. The disease is named after Thomas Addison, who first mentioned it in 1855.

What is yellow fever?

Yellow fever is a viral disease caused by mosquitoes. It damages many tissues of the human body, especially the liver causing intense jaundice.

What is cirrhosis?

It is a disease of liver in which tissues of liver become scarred and useless. It is mostly caused by excessive use of alcohol, a poor diet or inhaling poisonous fumes.

What is pulmonary tuberculosis?

It is a disease of lungs caused by bacteria called mycobacterium tuberculin. (Fig. 17.2)

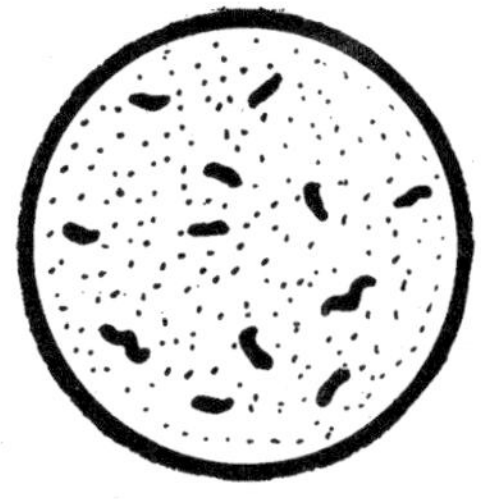

Fig. 17.2 Mycobacterium tuberculin

What is arteriosclerosis?

It is a diseased condition of the walls of the arteries, which become weakened, hard and twisted. In other words, arteriosclerosis is a narrowing of the wall of the blood vessels by thickening due to the deposition of fibro-lipid substances.

What causes scabies?

It is caused by parasitic insects.

What is arthritis?

It is a disease of joints. In this disease joints become swollen and crooked, and their movement becomes difficult. Sometimes it can cripple a person.

What is jaundice?

Jaundice is a condition characterized by the unusual presence of bile pigment in the blood. The bile produced in the liver passes into the blood instead of the intestines and because of this there is a yellowing of the skin and the whites of the eyes. (Fig. 17.3)

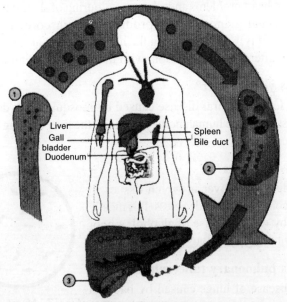

Fig. 17.3 Jaundice

What is the main cause of tooth decay?

Sugar in the diet is the most common factor of tooth decay.

What is botulism?

Botulism is the most dangerous type of food poisoning caused by the anaerobic *bacterium clostridium botulinum*.

What is the disease of rickets?

Rickets causes children's bones to become soft and deformed. It is caused by the lack of vitamin D and calcium in the diet.

What is an allergy?

A hypersensitivity to something in the environment which is harmless to most individuals. The offending substance or condition is called an allergen. The most common allergens include pollen, food, drugs, bacteria and house dust. However, some individuals have allergic reactions to cold, heat, or light, and even to emotional stress. The parts of the body in which allergic reactions most often come up are the skin, bronchial tube, nose and digestive system. (Fig. 17.4)

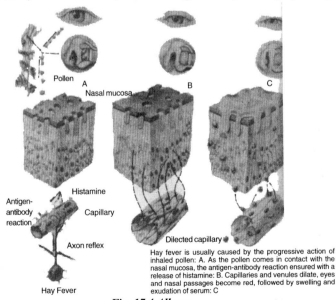

Hay fever is usually caused by the progressive action of inhaled pollen: A. As the pollen comes in contact with the nasal mucosa, the antigen-antibody reaction ensured with a release of histamine: B. Capillaries and venules dilate, eyes and nasal passages become red, followed by swelling and exudation of serum: C

Fig. 17.4 Allergy

What is a haemophilia?
Haemophilia is a hereditary disorder of blood coagulation in which the blood clots very slowly. If a person suffering from this disease gets any cut, it may prove fatal.

What is glaucoma?
Glaucoma is an eye disease which is caused by an increase in the amount of fluid inside the eyeball. It may cause blindness if not treated properly.

What is colour blindness?
Colour blindness is inability to distinguish colours. This is an inherited characteristic, and much more common in men than in women. Much of colour blindness is present for a limited number of colours, e.g., red and green. In extreme cases all colours appear grey.

Which vitamin's deficiency causes night-blindness?
Vitamin A.

What does the abbreviation AIDS stand for?
Acquired Immunity Deficiency Syndrome.

What causes AIDS?
It is caused by a virus, mainly by homosexuality, unprotected sexual contact, shared syringe among drug users and by infected blood products.

What is meningitis?
It is a disease caused by bacteria called *neisseria meningitis*. It attacks brain and spinal cord. If not treated immediately, the patient dies. (Fig. 17.5)

What is cancer?
Cancer is a disease in which certain group of cells start multiplying without control, and destroying healthy tissues.
It is also a widely-used term describing any form of malignant tumor.

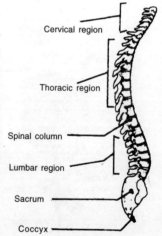

Fig. 17.5 Spinal cord

Is cancer comparatively a modern disease?
No. Evidences have been found in early Egyptian mummies.

How many type of cancers are there?
More than 100 type of cancers attack human bodies. The most common cancer sites in men are skin, lung, prostrate, large intestine, urinary tract, blood and lymph. Common cancer sites in women include breast, skin, large intestines, uterus, lung, blood and lymph.

What is meant by double pneumonia?
Pneumonia in both lungs. It is a bacterial infection of the lungs resulting inflammation and filling of the alvoli with pus and fluids.

What is the difference between obstetrics and gynaecology?
Obstetrics is the treatment of women during pregnancy, childbirth and the period immediately following, whereas gynaecology is a diagnosis and treatment of ailments concerning sexual and reproduction functions and disease of reproduction organs of females.

What does medical term 'abortion' mean?
The spontaneous or artificially induced expulsion from the mother's womb of a non-living embryo or child during the early stages of development. A spontaneous abortion is often called a miscarriage.

Why is an antidote administered?
To counteract the effect of poison.

Who is an anaconic?
A person who lacks the sense of smell.

What is the epiglottis?
A valve in the throat that covers the top of the windpipe (trachea) during swallowing.

Which diseases occur more commonly in advanced age?
The diseases which occur more frequently in advanced age are chiefly degenerative diseases: hardening of arteries; heart disease; malignant growths such as cancer; strokes; prostate trouble, senility. Degeneration means deterioration of tissues with loss of function and eventual destruction of the particular tissue cells.

18. MEDICINE AND MEDICAL ENGINEERING

What is medicine?
Medicine is the science of preventing, diagnosing and treating disease, both physical and mental.

What are the different systems of medicine?
There are mainly four systems of medicine: Allopathic, Ayurvedic, Homeopathic and Acupuncture.

What is acupuncture?
Acupuncture is a Chinese medical system in which small metal needles are inserted into the body at different points for relieving pain and curing various diseases. Sometimes it is employed as an alternative to anaesthesia. Experts know the specific points where needles are to be inserted. (Fig. 18.1)

Fig. 18.1 Acupuncture

What is internal medicine?
It is a branch of medicine that deals with the diagnosis and treatment of ailments caused by infections, systemic, metabolic, nutritional, physical and chemical agents. It covers the diseases of respiratory, kidney, cardiovascular (heart and blood vessels), endocrine, digestive and nervous systems.

What is chemotherapy?
Treatment of diseases with chemicals. It usually refers to treatment of cancer with cytotoxic and other drugs.

What are antibiotics?
Antibiotics are the special kind of medicines which kill micro-organisms causing diseases in our body. The first antibiotic was penicillin discovered in 1928. Today, we have large number of

antibiotics such as tetracyclin, chloromycetin, ampicillin, streptomycin etc.

What are sulpha drugs?

Sulpha drugs are the synthetic chemicals used for the treatment of bacterial diseases. Sulphonamide is one important sulpha drug. Some side effects may occur but generally these are outweighed by the benefits.

What are tranquilizers?

Tranquilizers are the drugs that have a soothing and calming effect, relieving stress and anxiety. There is however, a danger of dependence with long-term use.

What are analgesics?

Analgesics are drugs which relieve pain varying in potency from mild such as paracitamol and aspirin, to very strong e.g. pethedine and morphine.

What is nuclear medicine?

Nuclear medicine is that branch of medical sciences in which radio isotopes are used to treat certain diseases, such as iodine-131 is used to treat goitre or thyroid diseases and cobalt-60 is used to treat cancer.

What is a placebo?

Placebo is an inactive substance like sugar pill or any other 'make-believe' drug taken as medication, that nevertheless may help to relieve a condition. The change occurs because the patient expects some treatment and improvement that reflects his expectations. New drugs are tested and measured in trails against placebo response.

What is a vaccine?

It is a modified preparation of micro-organisms or of their toxins used for combating disease by inoculation.

Who developed the first small-pox vaccine?

Edward Jenner.

What are the uses of X-rays in medicine?

X-rays are used to detect certain diseases of human body, such as the fracture of a bone and lungs TB etc. They are very useful in therapy and diagnosis within medicine. They are also used to treat some type of cancers. (Fig. 18.3)

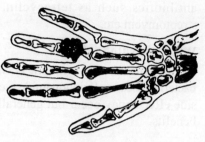

Fig. 18.3 X-ray photo

What is the difference between anaesthesia and hypesthesia?

Anaesthesia is the complete loss of the sense of pain or touch, whereas hypesthesia is the diminished capacity for sensation.

What is E.E.G. or electro encephalography?

It is a technique of recording the electrical activity of brain by means of attaching electrodes to the scalp to investigate some brain disorders.

What is E.C.G. or electro cardiography?

It is a technique of recording heart beat for investigating heart diseases. The patient is connected to the equipment by leads on the chest and legs and arms.

What is electro-retinogram?

The recording of potential changes produced by the eye when retina is exposed to a flash of light is called the electroretinogram. It is used to detect the diseases of retina. (Fig. 18.4)

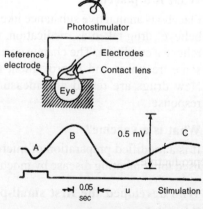

Fig. 18.4 E.R.G.

What is electromyogram?

Electromyogram is the recording of electric signals produced by human muscles. It gives valuable information regarding muscle disorders.

What is an auriscope?

It is an instrument used for examining ear disorders. (Fig. 18.5)

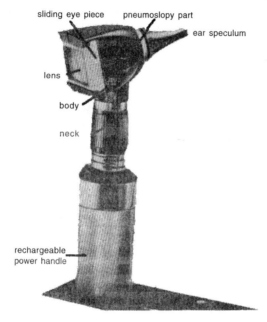

Fig. 18.5 Auriscope

What is a heart-lung machine?

It is a life-saving machine. The apparatus is used during heart surgery to take over the functions of the heart and lungs temporarily.

What is artificial insemination?

It is the method of injecting sperms of a prized animal into reproductive organ of the female animal.

What is an artificial pace-maker?

It is a device which is fitted on the chest of the patient to regularize the irregular heart beats. It is equipped with a battery which can last for years together.

What is a CAT scanner?

CAT scanner is a medical instrument used to detect the disease of brain, kidney, liver, abdomen, etc. by using soft X-rays. It is also used in archaeology to investigate mummies. The diagnosis is done by a computer. The word CAT stands for Computerized Axial Tomography. (Fig. 18.6)

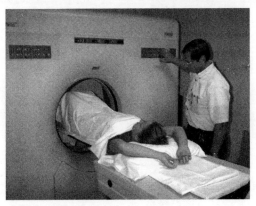

Fig. 18.6 CAT scanner

What is an endoscope?

An endoscope is an optical instrument used to examine the internal organs and tissues of the body.

What is ultrasound diagnosis?

To make use of ultrasonic waves for detecting certain diseases of the kidney, abdomen, etc. is called ultrasound diagnosis. Ultrasonic waves are sent through the body to take pictures of the insides of the body.

What is radiotheraphy?

Radiotheraphy is the use of X-rays or other radiations to cure diseases.

What is artificial kidney machine?

Artificial kidney machine is a kind of dialyser used to save patients suffering with kidney problems. The patient is plugged into the machine. This removes the waste products of blood by filtering it and returns the pure blood to the body.

What is a test tube baby?

A test tube baby is no different from any other baby. But instead of beginning to grow inside its mother's womb, it begins outside her body in a special machine. An ovum (egg) is taken from the mother and fertilised with the father's sperm in the machine which is called a test-tube, but is not really. The embargo thus developed is transplanted in the uterus of the mother.

Who was the first test tube baby of the world?

Louise Joy Brown was the world's first test tube baby. She was born to Lesley and Gilbert Brown of the U.K. on July 25, 1978.

Who is known as the father of medicine?

Hippocrates.

Who made the first heart plant?

Christiaan Bernard in 1967.

●●●

19. UNITS

What are the basic units in International System of Units?

There are seven basic units in metric system namely metre for length, kilogram for mass, second for time, ampere for electric current, kelvin for temperature, candela for luminous intensity, and mole is the unit for measuring the quantity of a substance involved in chemical reactions.

What is the unit of heat?

Calorie or joule. A calorie is defined as the quantity of heat required to raise the temperature of 1 gram of water through 1°C and 1 calorie = 4.1868 joules.

What is the unit of pressure?

Pascal. It is named after the French mathematician Blaise pascal or torr.

What is light year?

Light year is a unit which measures the distance between stars. It is defined as the distance travelled by light in vacuum in one year. It is equal to 58,80,030 million miles or 9.46 trillion km (Fig. 19.1). In this figure distances of moon, sun and alpha-centauri are shown in terms of light seconds, minutes and years.

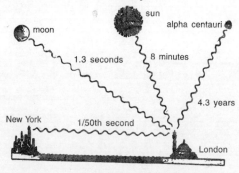

Fig. 19.1 Light year

What is an Angstrom unit?

It is mainly used to express the length of light waves. It is named after the Swedish Scientist A.J. Angstrom. An Angstrom unit is one-ten-millionth of a millimeter (10^{-7} mm).

How do we define astronomical unit?

An astronomical unit represents the mean distance between the sun and the earth. Its value is 149,600,000 km. One light year = 60,000 A.U.

What is the unit of atmospheric pressure?

Bar. One bar is equal to a pressure of 10^6 dynes per sq. cm.

In which unit do we measure the speed of ships?

Knots. A knot is the abbreviated term for one nautical mile which is equal to 6,080 ft. (Fig. 19.2)

Fig. 19.2 Knot

What are the units of frequency, force, work, power and energy?

The units of frequency, force, work, power and energy are Hertz, Newton, Joule, Watt and Joule respectively.

In which unit is the speed of supersonic planes measured?

Mach numbers. One Mach number is equal to the speed of sound, i.e., 340 metres per second at sea level. It is named after Austrian physicist Ernst Mach.

Which unit is used to measure the electrical resistance?
Ohm.

Name the three systems of measurements?
Three systems of measurements are MKS (Metre, Kilogram, Second), CGS (Centimetre, Gram, Second), and FPS (Foot, Pound, Second).

What are the units of force?
Newton is the unit of force in MKS system, while dyne is used in CGS system.

In which unit is electric charge measured?
Coulomb.

What is par-sec?
It is a unit of distance. It is equal to about 200,000 times the distance of the earth from the sun, or about three and one-fourth light years.

Which unit measures sound intensity?
Decibels.

Which unit measures wine?
Hogshead. One hogshead = 52.5 gallons.

Which unit measures water depths?
Fathom. One fathom = 6 ft. or 1.8 metres.

What for is barrel used?
Barrel is the unit of liquid capacity, which depends on the liquid being measured. For petroleum, a barrel contains 35.5 gallons or 159 litres; or a barrel of alcohol for example, contains 41.5 gallons or 189 litres.

How many sheets of paper are there in one ream?
500 or 516 (formerly 480).

Which unit is used to measure magnetic induction?
The SI unit of magnetic induction is Tesla which is equal to one Weber per square metre and equivalent to 10^4 gauss.

Which unit is used to measure electrical energy?
Kilowatt hour. 1 Kw hr = 3.6×10^6 Joules.

What for is horse power used?
Horse power is the unit of power. One horse power = 746 watts. It was first used by James Watt who compared the power of steam engine, with that of horses.

What is giga?
One giga is equal to 10^9, i.e., one thousand million or one billion. For example, one giga watt = 10^9 watts.

What is mega?
One mega = 10^6, i.e., one million.

Explain the terms terra, pico, nano and femoto.
One terra = 10^{12}, one pico = 10^{-12}, one nano = 10^{-9} and one femoto = 10^{-15}.

What for is fermi used?
Fermi is used to measure the radii of nuclei. One fermi = 10^{-15} m. The unit is named after Enrico Fermi.

What is photon?
Photon is a light particle whose energy is *hv*.

What is quart?
Unit of capacity equal to one quarter of a gallon.

What is 1 gallon?
8 pints.

What is 1 hectare?
10,000 m².

What is acre?
4,046.9 m².

What is 1lb?
454 gm.

●●●

20. SCIENCE LAWS AND EFFECTS

What is Doppler's effect?
It can be described as the apparent change in the frequency of sound; light or electro-magnetic waves caused by the motion of the source, observer or medium. It was given by Austrian physicist, Christian Doppler, in 1842. It is used to study the speed of stars.

What is piezoelectric effect?
When pressure is applied on certain crystals, such as Rochelle's salt, an electric voltage develops across the crystal. This is called piezoelectric effect. Piezoelectric crystal oscillators are used as frequency standards and for producing ultrasonic oscillations. (Fig. 20.1). If a voltage is applied to a piezoelectric crystal the crystal expands or contracts. This is called reverse piezoelectric effect.

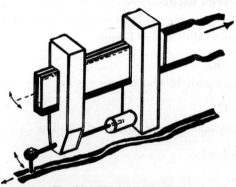

Fig. 20.1 Piezoelectric effect

What is Bernoulli's effect?
According to Bernoulli's effect, as the velocity of a liquid or gas increases, its pressure decreases so that the total energy remains constant. This effect is very useful in aerodynamics. The principle was named after Swiss mathematician and physicist Daniel Bernoulli.

What is photoelectric effect?

When light falls on the surface of certain substances such as nickel, tungsten, etc., electrons are emitted when it is struck by photons usually those of visible light or ultraviolet radiation. This is called photoelectric effect. (Fig. 20.2)

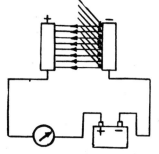

Fig. 20.2 Photoelectric effect

What is the law of floatation?

The law of floatation states that for the floatation of a body following conditions should be fulfilled:
 (i) The weight of the body should be equal to the weight of the water displaced.
 (ii) The centre of gravity of that body and that of liquid displaced should be in the same vertical line.

What is Newton's law of gravitation?

Newton's law of gravitation states that every mass in this universe attracts every other mass with a force, whose direction is that of the line joining the two and whose magnitude is directly as proportional to the product of the masses and inversely as proportional to the square of the distance between them.

If 'm_1' and 'm_2' are the two masses separated from each other by a distance 'r', the force will be $F = \dfrac{Gm_1 m_2}{r^2}$ (G = universal gravitation constant)

The value of $G = 6.6 \times 10^{-11}$ Nm/Kg². (Fig. 20.3)

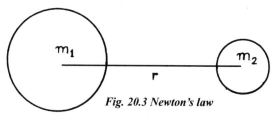

Fig. 20.3 Newton's law

State Newton's laws of motion?

Newton gave three laws of motion:
 (i) According to the first law, if a body is at rest or is in a state of

uniform motion in a straight line, it will remain in the same state unless it is disturbed by some external force to change that state.

(ii) The second law of motion states that the rate of change of momentum is directly proportional to the impressed force, and is always in the direction of the force.

(iii) The third law of motion states that to every action, there is an equal and opposite reaction.

What is Newton's law of cooling?

Newton's law of cooling states that the rate of loss of heat of a hot body is directly proportional to the difference of temperature between the body and the surroundings and is independent of the nature of the body.

What is Ohm's law?

Ohm's law states that the ratio of the potential difference between the ends of a conductor and the current flowing in the conductor is constant, e.g., for a potential difference of E volts and a current of 1 amperes, the resistance R, in Ohms is equal to E/1. The law was discovered by German physicist George Ohm in 1827.

What is the principle of conservation of energy?

The principle of conservation of energy states that in any system energy cannot be created or destroyed; the sum of mass and energy remains constant.

What is Snell's law?

Snell's law states that the ratio of the sine of angle of incidence to the sine of the angle of refraction remains constant for any two given media. The law is named after its discoverer, Dutch physicist Willerbrord Snell. Mathematically it can be represented as: $\frac{\sin i}{\sin r} = \mu$ (a constant). (Fig. 20.4)

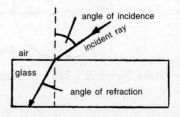

Fig. 20.4 Refraction

What is Raman Effect?

When light of one frequency is transmitted through a medium, other frequencies get added or subtracted from it. These are the characteristic frequencies of the material. This is known as Raman Effect. This is also the basis for Raman spectroscopy.

What is Boyle's law?

Boyle's law states that the volume of certain gas is inversely proportional to the pressure at a constant temperature. In other words, the product of pressure and volume remains constant provided the temperature is kept constant, i.e., $P \times V$ = a constant if T remains the same. Real gases, however, considerably deviate from this law.

What is Charle's law.

Charle's law states that the volume of a given mass of a gas at constant pressure is directly proportional to the absolute temperature i.e. $V \times T$.

What is Dulong and Petit's law?

Dulong and Petit's law states that the product of atomic weight and specific heat of solid elements is nearly equal to 6.4, i.e. atomic wt×sp heat = 6.4 approx.

What is Gay Lussac's law of combining volumes?

Gay Lussac's law of combining volumes states that gases react together in volumes which bear simple whole number ratios to one another and also to the volumes of the product, if all the volumes, being measured under similar conditions of temperature and pressure.

What is Archimedes Principle?

According to this principle, when a body is imersed in a fluid, the apparent loss in its weight is equal to the weight of the fluid displaced. It was discovered by Greek mathematician Archimedes.

State Avagadro's law?

Equal volumes of all gases under the similar conditions of pressure and temperature contain equal number of molecules. It was propounded by Italian physicist Amedeo Avagadro.

What are laws of reflection?

The angle of incidence is equal to the angle of reflection. The incident ray, reflected ray and normal ray, all lie in the same plane.

What is Bode's law?

It is an empirical law that gives approximate distances of the planets from the sun. It was given by Johann Bode. According to this law, the distance D in astronomical unit is given by $D = 0.4 + 0.3 \times 2^n$, where $n = 0, 1, 2...$

What is Graham's law of Diffusion?

According to this law, the rates of diffusion of gases are inversely proportional to the square roots of their densities under similar conditions of temperature and pressure.

What is the law of Definite Proportion?

The law of Definite Proportion states that a chemical compound is always found to be made up of the same elements combined together in the same ratio by weight.

What is the Seeback Effect?

When the ends of two wires made from dissimilar metals are joined together and the two junctions are kept at different temperatures, an electric current flows through the wires. This is known as the Seeback Effect after the German physicist, T.J. Seeback, who discovered it in 1921. This is also called the thermoelectric effect, and is the basis of thermocouple.

What is Pascal's law?

When pressure is applied anywhere in an enclosed liquid, it is transmitted equally in all directions. It was discovered by the French physicist, Blaise Pascal. (Fig. 20.5)

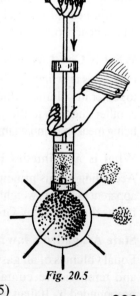

Fig. 20.5

21. INVENTIONS AND INVENTORS

What was the important invention of Alfred Nobel?
Alfred Nobel of Sweden invented dynamite in 1867. He earned from dynamite and left $9,000,000 to set up the prizes that now bear his name called Nobel Prizes. The interest that this money earns each year is given away as the awards.

What for is Ernest Lawrence famous?
Ernest Lawrence of the USA is famous for the discovery of cyclotron. It is a machine used to accelerate subatomic particles. It was discovered in 1930. The first successful model was built in 1931. (Fig. 21.1)

Fig. 21.1 Ernest Lawrence

Who invented photography?
Joseph Niepce of France. In 1826, he made the first successful permanent photograph.

Who invented compound microscope?
Zacharias Janssen of Holland in 1590.

What is the biggest contribution of Roentgen?
Wilhelm Roentgen Conrad of Germany discovered X-rays in 1895 for which he was awarded the first Nobel Prize in 1901.

Who discovered telephone?
Alexander Graham Bell of the USA in 1876. (Fig. 21.2)

Fig. 21.2 Telephone

Who discovered electron?

Sir J.J. Thomson of Britain, who is regarded as the founder of modern physics. Thomson was awarded the 1906 Nobel Prize for physics. (Fig. 21.3)

Who discovered Neutron?

James Chadwick.

Fig. 21.3 Sir J.J. Thomson

Who gave the Quantum Theory?

Max Plank in 1901.

Who gave the special theory of relativity?

Albert Einstein of Germany in 1905. (Fig. 21.4)

Who invented Xerography?

Chester Carlson of the USA in 1938.

Fig. 21.4 Albert Einstein

What are the contributions of Edison?

Thomas Alva Edison invented many things and probably the greatest of all time with over 1000 patents issued to his name which include automatic telegraph, electric bulb, phonograph, generator, etc. (Fig. 21.5)

What are the contributions of the following?
(a) Charles Darwin (b) Louis Pasteur (c) Hideki Yukawa.

(a) Charles Darwin propounded the theory of evolution.

(b) Louis Pasteur, a French chemist, discovered pasteurization. He also discovered an inoculation against dreaded disease of anthrax and hydrophobia or rabies.

Fig. 21.5 Thomas Alva Edison's electric bulb

(c) Hideki Yukawa of Japan discovered mesons. He won Nobel Prize in 1949.

Who discovered micro-processor?
Robert Noyce and Gordon Moore of the USA in 1971.

Who was the greatest ancient inventor?
Hero of Alexandria. He invented steam globe. (Fig. 21.6)

Who discovered refrigerator?
James Harrison and Alexander Catlin of the USA in 1850.

Fig. 21.6 Steam globe

Who invented stethoscope?
Dr. Reme Theophile Hyacinthe Leannec of England.

Why is Henri Becquerel well known?
Henri Becquerel, a French physicist, is well known for discovering radioactivity. Madam Curie worked with him.

What were the inventions of Leonardo da Vinci?
Leonardo da Vinci of Italy was not only an artist but also a scientist and engineer. He developed the models of cannon, modern helicopter, war machines, flying machine, etc. (Fig. 21.7)

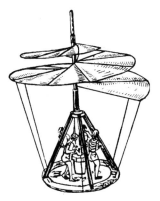

Fig. 21.7 Flying machine

What for is Dmitri Mendeleev known all over the world?
Dmitri Mendeleev, a Russian chemist, is known the world over as the founder of periodic table and for the development of petroleum and other industries in Russia.

Who invented steam engine?
A Scottish engineer, James Watt.

Who discovered the cause of malaria?
Ronald Ross of Britain. He was awarded Nobel Prize for medicine in 1902.

Who invented the thermometer?
The Italian scientist, Galileo in 1593.

Who correctly explained the role of oxygen in combustion?
Antoine Lavoisier (1773-1794) — a French born chemist.

Who made the first atom bomb?
Otto Hahn of German origin. He won the Nobel Prize for chemistry in 1944.

Who discovered radium?
Pierre Curie and Madam Marie Curie of France in 1898. It was obtained from pitchblende. It is used in medicine to destroy cancer tumours. (Fig. 21.8)

Who invented polaroid camera?
Edwin H. Land of the USA. The first polaroid camera was sold in 1948.

Fig. 21.8 Pierre Curie

What famous hypothesis was put forward by Nicholas Copernicus in 1543?
He gave the Heliocentric Theory. According to this theory the sun was the centre of universe, and not the earth.

Who designed the first practical typewriter and when?
Christopher Lathem Sholes and Carlos Glidden of the USA in 1868. Its patent was given to Remington company in 1873.

Who contributed greatly to the development of modern nuclear physics?
Niels Bohr (1885-1962) — the Danish physicist. He won the Nobel Prize for physics in 1922.

What for is Thomas Hunt Morgan famous?

Morgan (1886–1945), an American zoologist and geneticist, is famous worldwide for his work linking chromosomes and heredity. He was awarded Nobel Prize for this discovery in 1933.

●●●

22. IMPORTANT SCIENTIFIC INSTRUMENTS

What is an altimeter?
Altimeter is an instrument used in aircraft that measures altitude, or height above sea level. The common type is a form of aneroid barometer, which works by sensing the differences in air pressure at different altitudes.

Which instrument is used to measure electric currents in amperes?
Ammeter.

What is the use of binoculars?
Binoculars are used for seeing distant objects clearly. Binoculars consist of two telescopes containing lenses and prisms, which produce a stereoscopic effect as well as magnifying the image.

Which device measures time in ships?
Chronometer.

What is crescograph?
It is used to measure the growth of plants. It was invented by J.C. Bose.

What is an epidiascope?
It is an optical device used for projecting films as well as images of opaque objects on a screen.

Which instrument is used to detect the presence of electric charge?
Electroscope.

Which instrument is used to measure the depth of ocean?
Fathometer.

What is an optical microscope?
An optical microscope is used to see minute objects magnified. It consists of an objective and eyepiece which produce an enlarged image of the object.

What is a Geiger-Muller Counter?
This special type of instrument is used for detecting and for counting nuclear radiations and particles. (Fig. 22.1)

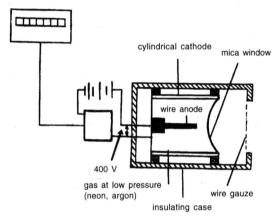

Fig. 22.1 Geiger counter

What is a voltmeter?
An instrument used to measure voltage.

Which device is used for measuring specific gravity of fluids?
Hydrometer.

Which instrument is used to measure the degree of moisture of the atmosphere?
Hygrometer.

What for is lactometer used?
It is used to test the purity of milk.

What is a gyroscope?

It is used to measure the rotation rate of ships and aeroplanes. It is a type of spinning-wheel fixed to the axle. Applications of gyroscope principle include the gyrocompass, the gyropilot for automatic steering and gyro-directed torpedoes. (Fig. 22.2)

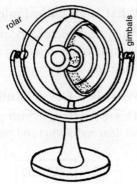

Fig. 22.2 Gyroscope

What for is tachometer used?

It measures the speed of running water or any flowing fluid. It also indicates the revolutions per minute of an engine, especially an automobile engine.

What is a periscope?

It is an optical instrument designed for observation from a concealed position such as from a submerged submarine. In its basic form, it consists of a tube with parallel mirrors at each end, inclined at 45° to its axis. (Fig. 22.3)

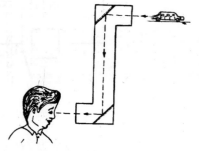

Fig. 22.3 Periscope

Which instrument is used to measure viscosity of fluids?
Viscometer.

What is the use of odometer?
An odometer records the number of revolutions of a wheel and thus indicates the distance travelled by a bus, motorcycle, car or a truck.

What is a pyrometer?
It measures high temperatures.

Which instrument measures the intensity of light?
Photometer.

What is an electron microscope?

An electron microscope manipulates the beam of electrons to achieve magnification. It uses magnetic coils rather than glass lenses to bend the electron beam and form enlarged images. It can magnify a minute object hundreds of thousands of times. (Fig. 22.4)

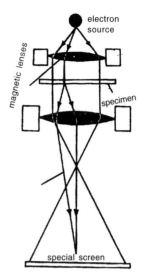

Fig. 22.4 Microscope

What is the use of radiometer?

A radiometer is used to measure the radiant energy, as well as intensity of light.

Which instrument is used to measure sugar concentration in a solution?

Saccharimeter.

Which instrument is used to measure blood pressure?

Sphygmomanometer.

Which instruments are used in weather forecasting?

Various instruments are used in weather forecasting, such as maximum and minimum thermometer, rain gauge, barograph, anemometer, wind vane, etc. Maximum and minimum thermometer is used to record the highest and the lowest temperature of the day. Rain gauge measures the amount of rainfall. Barograph continually measures the air pressure, and anemometer measures the velocity of wind. A wind vane detects the direction of wind.

What is a spectrometer?

It is an instrument used to measure the refractive index and the angles of a prism. It is also used to study spectrum.

What is a theodolite?

It is used for measuring horizontal and vertical angles by means of a telescope attachment. It is used in surveying.

What is a kymograph?

It is used to record graphically blood pressure and heart beat. It is also used to study lungs.

Which instrument is used to detect strength and direction of electric current?

Galvanometer.

Which instrument is used to measure rainfall?

Rain gauge. (Fig. 22.5)

Why is the clinical thermometer made oval instead of round?

So that it may act as a magnifying glass, thus making it easier to read.

Fig. 22.5 Rain gauge

What is a telescope?

It is an optical instrument used for making distant objects appear nearer and larger.

Who discovered thermos flask?

James Dewar.

What is a stroboscope?

A stroboscope studies the rate at which the objects rotate or vibrate.

Which instrument is used to find the direction in ships and aeroplanes?

Magnetic compass. (Fig. 22.6)

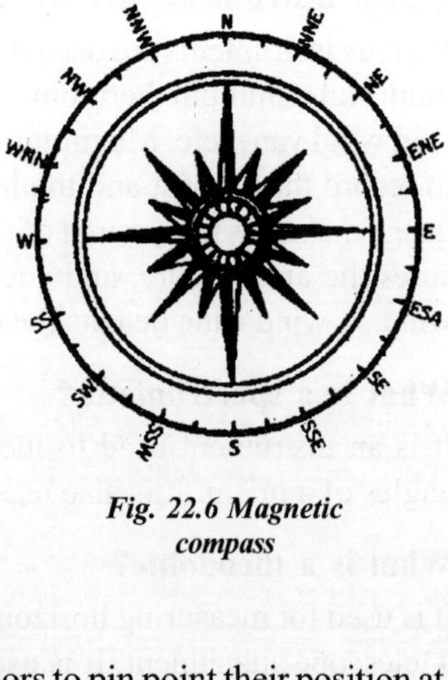

Fig. 22.6 Magnetic compass

What is the navigational sextant?

It is an invaluable instrument for sailors to pin point their position at sea.

What is a reflecting telescope?

It uses a system of concave mirrors to collect and focus incoming light. (Fig. 22.7)

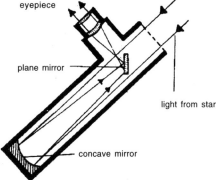

Fig. 22.7 Newtonian telescope

What is a radio telescope?

This is an astronomical instrument which receives, amplifies and measures radio waves from outer space with the help of parabolic reflectors. These are further analysed in computers for studies.

What is dip circle?

An instrument which measures the angle of dip. (Fig. 22.8)

What are refracting-type telescopes?

A refracting type telescope uses a set of lenses to focus light. One set of lenses serves the purpose of objective lens and the other as an eyepiece.

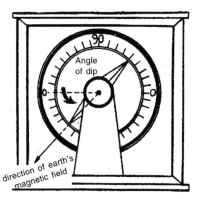

Fig. 22.8 Dip circle

●●●

23. SCIENTIFIC ACHIEVEMENTS OF INDIA

Which was India's first satellite?
Aryabhatta was India's first satellite. It was launched on April 19, 1975 from Russia. It weighed 360 kg and was 16 m high with 26 faces. Its objective was to conduct experiments in X-rays, astronomy and physics. (Fig. 23.1)

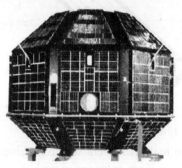

Fig. 23.1 Aryabhatta

When was Bhaskara launched?
Bhaskara, India's second satellite, was launched on June 7, 1979 from Russia.

When was Bhaskara-II launched?
Bhaskara-II, an earth observation satellite, was launched on November 20, 1981 from Russia.

What do you know about INSAT-IA?
INSAT-1A was India's first operational multi-purpose and unique domestic satellite. It was meant to enhance the communicational, meteorological and television relay and radio broadcasting capabilities. It was launched on April 10, 1982 from Cape Canaveral (U.S.A.).

When was INSAT-1B launched?
India's multipurpose domestic satellite, INSAT-1B, was launched

on August 30, 1983 by a space shuttle. It is functioning successfully even today.

What are the latest trends of the Indian space endeavour?

Indian space endeavour is gaining international recognition and marketed worldwide, and saw three satellites. INSAT-2C, IRS-1C and IRS-P3 (Indian Remote Sensing) commissioned during 1995-1996. INSAT 2-C was launched in December, 1995, while IRS-P3 in March 1996. The next satellite INSAT-2D is in the final stages of fabrication. Work on INSAT-2E also commenced in 1996 and was launched in 1998.

Which was the first satellite-launch vehicle of India?

SLV-3 was India's first satellite launch vehicle. It put Rohini satellite into orbit on July 18, 1980. It was fabricated at the Vikram Sarabhai Space Centre, Thiruvananthapuram. (Fig. 23.2)

What do you know about India's Geosynchronous Satellite Launch Vehicle (GSLV) programme?

The major milestone was crossed in GSLV programme for launch of INSAT class satellites when one-tonne cryogenic engine development was completed in 1996. First GSLV flight has been tentatively fixed for 1998.

Fig. 23.2 SLV-3

Who was the first Indian to enter into space?

Sq. Ldr. Rakesh Sharma was the first Indian to enter into space aboard Soyuz T-II Russian spaceship. He went into space on 3rd April, 1984, along with two Soviet cosmonauts.

Name the places where atomic power stations have been established in India?

India's atomic-power plants are: (i) Rana Pratap Sagar atomic power plant (Rajasthan), (ii) Tarapur power plant, Tarapur (iii) Kalpakkam atomic power plant, Kalpakkam (iv) Narora atomic power station, Narora.

Which are the places where Indian space centres have been established?

(1) Vikram Sarabhai Space Centre, Thiruvananthapuram;
(2) SHAR Centre, Sriharikota;
(3) ISRO Satellite Centre, Bangalore.
(4) Auxiliary Propulsion System Unit, Bangalore.
(5) Space Applications Centre, Ahmedabad.
(6) Development and Educational Communication, Ahmedabad and ISRO Telemetry Traking and Command Network, Bangalore.

What are the five nuclear reactors working at BARC, Trombay?

These are Apsara, Zerlina, Purnima, Circus and Dhruva.

When was Apsara commissioned and what are its functions?

Apsara was commissioned on August 4, 1956. It is one megawatt swimming pool-type reactor. It produces radio isotopes to irradiate biological samples etc.

What is India's biggest achievement in nuclear industry?

Fast Breeder Test Reactor at Kalpakkam marks a milestone in the building of advanced and indigenous nuclear industry. It was commissioned in 1985. It is a 13-megawatt reactor. It uses a new type of fuel, a plutonium carbide combination. (Fig. 23.3)

Fig. 23.3 Fast breeder reactor

Where are India's five famous observatories?

India's five famous observatories are at Kodaikanal, Hyderabad, Ooty, Nainital and Kavalur. (Fig. 23.4)

What for is Purnima being used?

Purnima (Plutonium Reactor for Neutronic Investigation in Multiplying Assemblies) became critical on May 22, 1972. It is a zero energy fast nuclear reactor.

Fig. 23.4 Observatory

When did Dhruva reactor become critical?

Dhruva reactor at Trombay became critical in 1985. It is a high-power reactor and its main functions are isotopes production and fuel material testing. It is also being used for basic research in physics, chemistry and biology.

Where is India's famous Radio Observatory?

It is in Udagamandalam (Tamil Nadu). It has a parabolic cylinder (530 m NS × 30 m EW) covering the approximate effective area of 8.0×10^3 sq.m.

Where is the largest reflecting telescope of Asia?

It is in the Kavalur Observatory (India). It is a 93-inch (236 cm) reflecting telescope and has been developed indigenously. (Fig. 23.5)

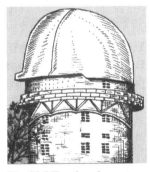

Fig. 23.5 Kavalur observatory

What is GMRT?

GMRT stands for Giant Metrewave Radio Telescope which is being set at Podar about 80 km. north of Pune, India. It consists of 3 steerable, parabolic dishes, each 45 m in diameter. It will help in study of Supernovae, Quasars, Pulsars and origin and evolution of the universe.

●●●

24. FAMOUS SCIENTISTS OF INDIA

Who was Aryabhatta?
Aryabhatta (born at Kusumapura near Patna in 476 A.D. and died in 550 A.D.) was a well known Indian astronomer and mathematician. He adorned the court of Chandragupta Vikramaditya. He wrote a book on mathematics called *Aryabhatta* in couplets.

Who was India's well known palaeobotanist?
Birbal Sahni (1891-1949) was India's famous palaeobotanist. He has been very famous for the studies of the Gondwana Flora and the problem of the age of the Saline Series of the salt range.

Who founded the Indian Chemical Society?
Acharya P.C. Ray (1861-1944) was the founder of the Indian Chemical Society and the Bengal Chemical and Pharmaceutical Works Ltd. He was the author of *Hindu Chemistry*. His work on nitrates is well-known.

Which Indian scientist carried out important studies on Relativity and Cosmology?
Prof. V.V. Narlikar (1908-1991). His significant contributions include generalization of Bode's Law of derivation of the highest atomic number in Eddington's Fundamental Theory and studies of gravitational space-time matrices.

Which India born scientist was awarded the Nobel prize for astrophysics?
Prof. S. Chandrasekhar was awarded the nobel prize for physics in 1983 for his outstanding researches in the field of astronomy. He was an India born scientist but he acquired American citizenship in 1953. He passed away in the year 1995.

Give a brief account of Dr. Vikram Sarabhai.

Dr. Vikram Sarabhai (born on August 12, 1919, at Ahmedabad and died on December 30, 1971) is well known in the field of cosmic rays. He gave new dimensions to India in the field of nuclear power and space programmes. Dr. Sarabhai was the brain behind the formation of the Physical Research Laboratory, virtually the cradle of Indian space programme. He was a big industrialist also. He was the recipient of Bhatnagar Memorial Award and Padma Bhushan. His contributions as Chairman of the Atomic Energy Commission are outstanding. (Fig. 24.1)

Fig. 24.1 Dr. Vikram Sarabhai

Who is India's well known aereospace scientist?

Professor Roddam Narasimha. He has carried out important studies on information of fluids i.e. liquids and gases. He has also developed and improved computer simulation methods for the study of aerospace problems. He was elected FRS in 1992. (Fig. 24.2)

Fig. 24.2 Roddam Narasimha

What for is Dr. Raja Ramanna well known?

Dr. Raja Ramanna is well known all over the world as a nuclear physicist. He was associated with India's first nuclear blast at Pokhran on 18th May, 1974.

Who did Asia's first heart transplant operation?

Dr. P.K. Sen, a well known Indian surgeon, performed Asia's first heart transplant operation in Mumbai.

What for is Dr. Jagjit Singh well known?

Dr. Jagjit Singh is well known as India's popular science writer and a Kalinga Prize laureate.

Who put India first on the science map of the world?

Sir Jagadish Chandra Bose, born on November 30, 1858, and died on Nov. 23, 1937, was the first person to put India on the science map of the world by his inventions. He developed an instrument called crescograph to detect the minute responses of living organisms, especially plants. He showed that plants respond to light rays and wireless waves. He wrote two world-famous books: *Response in the Living and Non-living* (1902) and *The Nervous Mechanism of Plants* (1926). (Fig. 24.3)

Fig. 24.3 Sir Jagdish Chandra Bose

What are the contributions of H.J. Bhabha?

Dr. Homi Jehangir Bhabha (born on October 30, 1909, in Mumbai and died on January 24, 1966, (in an air crash), was not only an eminent cosmic rays scientist, but also a skilled administrator. He did significant work in identifying the elementary particles called mesons. He was the first chairman of the Atomic Energy Commission of India. The Bhabha Atomic Research Centre has been named after him. (Fig. 24.4)

Fig. 24.4 H.J. Bhabha

Who was Meghnad Saha?

Meghnad Saha (born on October 6, 1893, at Dacca and died on Feb. 16, 1956, in New Delhi) was known as Palit Prof. of Physics. He is well known for his researches in nuclear physics, cosmic rays and spectrum analysis. His theory of thermal ionization brought him world fame. *History of Hindu Science* is one of his famous publications.

What for is Prof. E.C.G. Sudarshan famous?

Prof. Sudarshan is famous for advancing the theory of particles-tachyons which could be moving faster than light. Presently, he is the director at the Centre for Practical Theory, University of Texas, USA.

Who was the first Director-General of CSIR?

Dr. Shanti Swarup Bhatnagar (born on February 21, 1894, at Bhera in Punjab and died on January 1, 1955) was the first Director-General of CSIR. His work on magneto-chemistry is well known the world over. He became FRS in 1943. He opened a chain of National Research Laboratories in India. The S.S. Bhatnagar Memorial Award was instituted in his honour in 1958. This award is given for the outstanding contributions in physics, chemistry, biology, engineering, medicine and mathematics to Indian scientists. (Fig. 24.5)

Fig. 24.5 Dr. Shanti Swarup Bhatnagar

Which Indian scientist is called the 'Father of Indian Green Revolution'?

Dr. M.S. Swaminathan. He ushered in the first Green Revolution in India. This Indian agronomist developed high-yielding strains of wheat and rice. He also introduced methods and techniques to raise agricultural production. He is a FRS and was Director-General of International Rice Research Institute, Manila, Philippines, between 1982 and 1988.

Which India born scientist is awarded the Nobel Prize for physiology and medicine?

Dr. Hargobind Khorana, now an American citizen, is a world renowned biochemist. He was born on January 2, 1922, at Rajpur in Punjab (now in Pakistan). He developed the method of synthesis of DNA and RNA for which he was awarded the Nobel Prize in 1968 in physiology and medicine along with M.W. Norenberg and R.W. Holley. He visited India in 1969. He was awarded Padma Bhushan.

What for did Sir C.V. Raman become a world famous scientist?

Sir C.V. Raman (born on November 7, 1888, at Trichinapoly and died on November 21, 1970, at Bangalore) is well known for his discovery known as the 'Raman Effect'. He was awarded the Nobel Prize for physics in 1930 for this outstanding discovery. He was also the recipient of the Lenin Peace Prize of 1958. Besides the Raman Effect, he gave the theory of the blue colour of ocean and theory of musical instruments. He was the founder of the Raman Research Institute. (Fig. 24.6)

Fig. 24.6 Sir C.V. Raman

Give a brief account of the work of Srinivasa Ramanujan.

Srinivasa Ramanujan (born on Dec. 22, 1887, at Erode and died on April 26, 1920, at Kambakonam) was one of the greatest mathematicians of modern times. He made significant contributions to the theory of numbers, theory of partitions and theory of continued fractions. (Fig. 24.7)

Fig. 24.7 Srinivasa Ramanujan

Who developed the Bose-Einstein statistics?

Satyendra Nath Bose is well known for the Bose-Einstein statistics. It is a new type of quantum statistics which was developed by him along with Albert Einstein. The particles which obey this statistics are called Bosons after him. He was born in Kolkata on January 1, 1894.

Fig. 24.8 Satyendra Nath Bose

He was one of the outstanding scientists of India. In 1958, he became the Fellow of Royal Society, London. The same year he was awarded Padma Vibhushan and was declared a National Professor. He died on February 4, 1974, due to heart attack. (Fig. 24.8)

Which Indian scientist developed the concept of white holes?

Indian Physicist Professor Jayant Vishnu Narlikar proposed a Theory of Gravitation in collaboration with Sir Fred Hoyle and developed the concept of white holes as sources of energy in cosmic object. He is also a popular science writer of eminence. (Fig. 24.9)

Fig. 24.9 Prof. Jayant Vishnu Narlikar

Which Indian scientist has carried out important work on Ionosphere?

Dr. A.P. Mitra. He has carried out his research on Ionosphere through ground-based and space technique. He also made important contributions to the study of Global Warming. He was elected FRS in 1989. (Fig. 24.10)

Fig. 24.10 Dr. A.P. Mitra

Who was Charaka?

Charaka was an ancient physician. He has written the famous book entitled *Charaka Samhita.*

Who was Brahmagupta?

Brahmagupta was a famous mathematician of Gujarat, born in 598 A.D. He made the rules of operation for zero.

What for has Dr. A.P.J. Abdul Kalam been famous?

Dr. Kalam, the President of India, has been famous for Missile Technology in India. Dr. A.P.J. Abdul Kalam made significant contributions in the field of missiles.

●●●

25. DOMESTIC APPLIANCES

How does a hair drier work?
A hair drier consists of a plastic cover. Inside this cover a coil of nichrome wire and a small fan are fitted. It has two switches, one for putting the fan on and other for passing the current through nichrome wire. When one switch is put on, electric fan starts running and cold air comes out. When the other switch is put on, the nichrome wire gets heated up and, as a result, the wind blown by the fan gets heated up and warm air comes out which is used for drying hair. (Fig. 25.1)

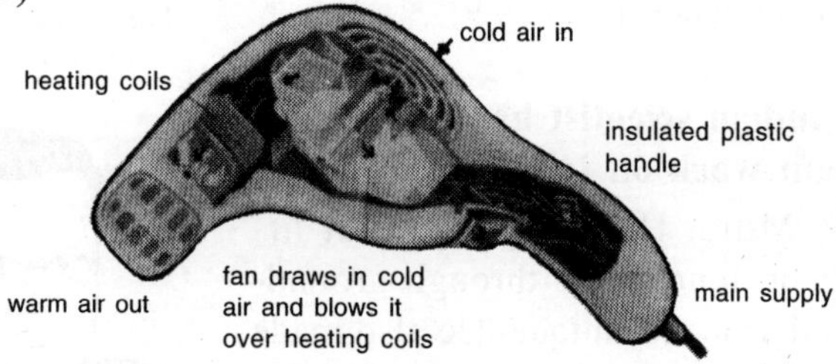

Fig. 25.1 Hair drier

How do electrically-heated appliances work?
All electrically-heated appliances, such as heater, electric iron, electric kettle, etc. are based on heating effects of electric current, i.e., when electric current is passed through a wire it gets heated up. In all appliances electric current passes through nichrome wire.

How does an electric fan work?
An electric fan consists of a motor. The armature of the motor is connected with a metal shaft. At the other end of the shaft, three metal blades are attached. As soon as the motor runs, these blades start rotating and produce cold air.

How does a refrigerator work?
A refrigerator is a device used for keeping food and other substances cool and fresh. It works on a cycle of vaporization and compression

of a volatile fluid usually feron gas. The liquid feron gas expands through a valve and evaporates, extracting heat from the contents of refrigerator. The vapour thus formed is compressed by a compressor and passes through a condenser whereupon it loses heat and liquefies. It is then recycled. (Fig. 25.2)

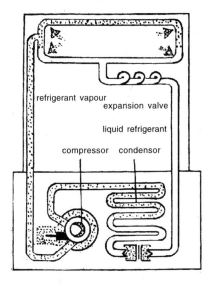

Fig. 25.2 Features of a refrigerator

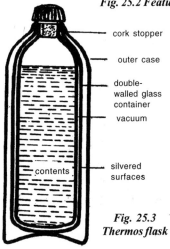

Fig. 25.3 Thermos flask

How does a thermos work?

A thermos flask consists of a double-walled glass bottle in which vacuum is created. The walls are coated with silver on both sides. The vacuum prevents heat transfer by convection, glass prevents heat transfer by conduction, and silvering on the walls prevents heat transfer by radiation. Therefore, hot liquids remain hot, and cold things remain cold for a long time in it. (Fig. 25.3)

How does a pressure cooker work?

A pressure cooker is a saucepan with an airtight lid in which food can be cooked quickly. Under pressure the water inside it boils at a much higher temperature than usual. As a result, food cooks more quickly. The pressure can be regulated by means of a weighted safety valve on the lid.

How do mixer and grinder work?

A mixer and grinder are fitted with a high speed electric motor. It is fitted with blades. As soon as the motor runs, the blades revolve. The rotating blades mince the food material into small pieces.

How does a vacuum cleaner work?

A vacuum cleaner is an electrically-operated machine used to clean carpets, rooms, floors, etc. It consists of an electric motor to which a small fan is attached. The fan rotates at a very high speed. It sucks the air in from one side and throws it to the other. A cloth-bag is fitted in the machine to collect the dirt.

How does an air-conditioner work?

An air-conditioner consists of a compressor and a cooling liquid, like feron gas. Cooling liquid evaporates in the cooling coil. The vapour is then carried to the electrically-operated compressor. It then goes to the condenser where it is cooled by air or water as it passes through the radiator. Here vapour changes into liquid giving up heat in the process. A fan helps to send fresh air into the room and thus keeps the room temperature at the desired level. (Fig. 25.4)

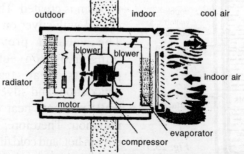

Fig. 25.4 Mechanism of a window-type air-conditioner

What is an electric blanket?

An electric blanket consists of heating wire and a thermostat. As soon as the current flows, the wires heat up. The temperature reaches a certain limit and thermostat cuts off the electric current. It is very useful in cold countries. (Fig. 25.5)

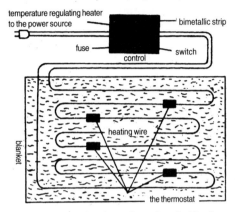

Fig. 25.5 Mechanism of an electric blanket

How does a toaster work?

The toaster is an appliance for toasting bread. An automatic toaster is made up essentially of a heater, a bimetallic thermostat and a spring. When the bread is toasted, it is raised up by spring mechanism, and electric current is shut off.

How does a gas lighter work?

In modern gas lighters, gas (butane, derived from natural gas) is pressurized in a tiny gas tank inside the cylinder. When the lighter is used, the gas shoots out of a small nozzle at the top where it is ignited by sparking device.

How does a microwave oven work?

Microwave ovens cook by means of radiation. The radiation is generated by a tube called magnetron. It makes the food molecules vibrate, creating heat by the internal friction that results from collision of the molecules.

●●●

26. EVERYDAY SCIENCE

At what temperature do Centigrade and Fahrenheit thermometers read the same?

At −40°.

Why is it dangerous to touch a live electric wire with bare hands?

When a live electric wire is touched with bare feet and hands, the current passes through the body and gives a shock which can be fatal.

Why does steam produce more severe burns than boiling water?

Steam contains 540 cal/gm of heat more than boiling water, hence it causes more severe burns.

Why is some space always left between two rail joints?

Some space is left between two rail joints because during summer the railway line expands due to heat. And if no space is left the railway track will bend causing an accident.

Why are fuse wires provided in electrical installations?

Fuse wire is a high resistance low melting wire. When a high electric current flows through this wire, it melts away avoiding the damage of the main installations. (Fig. 26.1)

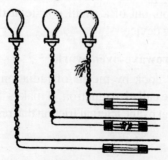

Fig. 26.1 Fuses

Why is sodium kept in kerosene, and phosphorus in water?
Sodium reacts with water to give sodium hydroxide and hydrogen. It starts burning. It is kept in kerosene to avoid its contact with water. Phosphorus is kept in water so that it remains cool. If placed in air, it can immediately catch fire.

How does a soda water straw work?
As a matter of fact we suck out air and to take its place, soda water rushes through the straw to the mouth.

Melting of ice wrapped in a blanket does not take place quickly, why?
A blanket is a bad conductor of heat, it cuts off the path of the heat rays. That is why the ice wrapped in a blanket does not melt quickly.

One leans forward while climbing a hill, why?
While leaning forward, the centre of gravity of the body also shifts forward, which helps climbing.

An electric bulb makes a bang when it breaks, why?
There is a partial vacuum in a bulb. As soon as it breaks, air rushes into it to fill the vacuum, thus producing sound or a bang.

Why are white clothes more comfortable in summer than dark colour ones?
White clothes absorb less heat than coloured ones. So one feels more comfortable in white clothes in summer than in coloured ones.

Why does blotting paper absorb ink?
The blotting paper is porous. These pores act as capillaries. The ink is sucked in these capillaries by surface tension, and hence the blotting paper absorbs ink.

Why does a thick glass tumbler generally break when hot liquid is poured in it?
When hot liquid is poured, the thick glass tumbler breaks because the inner surface of the glass expands due to heat but the outer

surface remains comparatively cool. Due to this unequal expansion, the thick glass tumbler breaks.

Why do we see a flash of lightning before hearing the thunder sound?

As we know, the speed of light is much faster than sound. Therefore, flash of lightning is seen before the sound of the thunder is heard.

How is milk converted into curd?

Milk contains a protein called casein. When curd is added to milk, the lactic acid producing bacteria present in it causes coagulation of casein and thus converts milk into curd.

Why does the rising and setting sun appear red?

The distance travelled by the sun rays in the atmosphere in the morning and evening is about fifty times more than the noon time. The dust, smoke and water vapours present in the atmosphere scatter different colours of sunlight differently. Violet is scattered the most, while red is scattered the least. Therefore, the amount of red light which reaches our eyes is maximum in the morning and evening. As a result, the rising and setting sun appears red.

Why does ice float on water?

Ice is lighter than water because when water gets converted into ice its volume increases. Nine volumes of water become 10 volumes of ice, i.e. ice is $9/10$ parts lighter than water. That is why it floats on water; its 9 parts remain under water and one part above the surface of water. (Fig. 26.2)

Why do we perspire on a hot day?

When the body temperature becomes elevated, the sweat glands inside the body are stimulated to secrete. This is nature's process to provide surface

Fig. 26.2 Ice pieces in the water

water for the evaporative cooling of body under heat stress. In cool environment, the glands remain entirely at rest.

How is rainbow formed?

A rainbow is an arc of seven colours: violet, indigo, blue, green, yellow, orange and red (vibgyor). It is produced when the sun rays fall on raindrops suspended in the air after rains. These rain droplets disperse the sun rays producing a band of seven colours. The seven colours of rainbow always appear in one and the same order. (Fig. 26.3)

Fig. 26.3 Rainbow

Why does an iron nail float on mercury but sink in water?

An iron nail floats on mercury because the weight of mercury is displaced by it, and hence the upward thrust becomes more than its weight. The density of mercury and iron is 13.6 gm per cc and 7.6 gm per c.c., respectively. The density of water being 1 gm/c.c., the upward thrust is much less, and hence the iron nail floats on mercury.

Why does sky appear blue?

The blue colour of the sky is due to the scattering of light by dust particles or air molecules. This scattering is inversely proportional to the fourth power of wavelength. Consequently, the shorter wavelengths are profusely scattered and when we look at a portion of the sky away from the sun, we receive these scattered and re-scattered light which is rich in blue, i.e., of short wavelength.

Why are we not hurt while cutting our nails?

The tip of our nails is made up of dead protein, called keratin. They have no connection with blood vessels or the cartilage. So the nerve system remains unaffected, and cutting of nails does not injure us.

Why does a red neck-tie appear black when seen under a blue mercury lamp?

During day light, the red neck-tie appears red because it absorbs all colours except red and reflects it. But when blue light from a mercury vapour lamp falls on it, it is wholly absorbed and no light is reflected, as the only component which it is capable of reflecting is missing. So the neck-tie appears black.

Why is water cold in an earthen pot in summer season than during rains?

An earthen pot has small pores from which evaporation of water takes place as air is dry and hot in summer, thus causing cooling of water. But during rains the air is already moist and cool. Therefore, evaporation does not take place quickly and water in the earthen pot does not cool. (Fig. 26.4)

Fig. 26.4 Pitcher

Why are we advised to empty ink from our fountain pen before aboarding an aeroplane?

As we aboard an aeroplane the air becomes rarer and the pressure of the atmosphere decreases. Hence, volume of air inside the fountain pen increases and the ink is pushed out, thus spoiling our clothes. And hence the advice.

Why does a straight stick partly immersed in water appear bent?

A straight stick partly immersed in water appears bent on account of the phenomenon of refraction. Accordingly, rays of light coming from the immersed portion of the stick bend away from the normal on emerging at the surface of water.

Why is it more difficult to breathe on mountains than on plains?

While breathing, a definite supply of oxygen is needed essentially. The density of air on mountains is less than on plains. Thus the oxygen content on mountains decreases. That's why it is more difficult to breathe on mountains than on plains.

Why are cloudy nights warmer than clear nights?

Cloudy nights are warmer than clear nights because clouds prevent radiation of heat from the ground and air.

Why are mountains cooler than plains?

Mountains are cooler than plains because:
(i) The air of mountains absorbs less heat than the air on plains;
(ii) The heat absorbed during the day on mountains radiates very quickly at night owing to the rarity of air and nights on mountains are cooler; and
(iii) Due to uneven surface on mountains, the major portion always remains in the shade. The sun does not heat much of the land which may heat the air.

Which is the smallest TV set?

Today Japan is marketing pocket TV sets. The screen of these TV sets measures 5 cm diagonally.

What is the full form of BASIC?

BASIC stands for Beginner's All Purpose Symbolic Instruction Code.

What for does ALGOL stand?

ALGOL stands for Algorithmic Language.

What is LOGO?

LOGO is a special purpose language and is used by children for playing games.

How is cloth woven?

Cloth is woven by a loom.

●●●

27. MISCELLANEOUS

Who gave theory of relativity?
Albert Einstein.

What are the two main theories of relativity?
Special theory of relativity and general theory of relativity are the two main theories of relativity. Special theory deals with the relativistic nature of mass, space and time; while general theory deals with the effects of acceleration and gravity.

What happens to the mass of a body when its speed becomes very high?
The mass of the body increases as its velocity increases.

What happens to the length of a body when its speed increases?
The length of the body decreases as its speed increases.

What are the different types of pollutions?
Usually there are four types of pollutions:
 (i) air pollution caused by smoke of automobiles and factories;
 (ii) water pollution caused by wastes from factories, homes and animals;
 (iii) land pollution caused by pesticides, empty cans, wrap papers, scrap, etc.; and
 (iv) noise pollution caused by the sound of aeroplanes, cars, radio, TV, etc.

What is antimatter?
It is believed to be made up of antiprotons and antineutrons with positrons orbiting round the nuclei.

What is absorption?
Absorption in chemistry means taking up of a gas by a solid or liquid or the take up of a liquid by a solid. In physics, it means the

conversion of the energy of EME into other forms of energy on passing through a medium.

What is sonic boom?

A noise like a thunderclap caused by the shock waves produced when an aircraft travels with supersonic speed.

How does a siren work?

Siren is a device that produces a piercing sound, used as a warning signal. It consists of a perforated disc or cylinder on to which jets of air or steam are blown. When the disc or cylinder is rapidly rotated, the jets are interrupted, generating sound vibrations of definite pitch.

What is inertia?

The resistance offered by an object to a force applied to it is called inertia.

What is momentum?

The momentum of a moving body is the function of its mass and velocity. Numerically momentum = mass × velocity.

Why does a body weigh more at Poles than at the Equator?

The value of 'g' is the maximum at Poles and least at the Equator. So a body weighs more at Poles than at the Equator.

If a gas is allowed to expand, will its temperature rise or fall?

Its temperature will fall.

What is a lever?

A lever is a machine by which a load can be overcome by a lesser effort. It is a rod or a bar which moves freely on a pivot called fulcrum. (Fig. 27.1)

Fig. 27.1 The lever

Mention the use of a screw.

A thread of a screw is a kind of inclined plane which winds around a core. Screws are used for lifting jacks and presses.

What is the speed of sound in air?

The speed of sound in air is 340 m/sec.

In which material does sound travel the fastest?

Sound travels the fastest in solids next in liquids, and the least in gases.

Define atomic number.

The number of protons in an atom is known as the atomic number of that element.

What is mass number?

The total number of protons and neutrons in an atom is called the mass number.

Mention the properties of proton and neutron.

Proton is a positively charged particle while neutron is neutral in nature. The masses of the proton and neutron are almost equal. Both these particles constitute the nucleus of the atom.

Name the defects of human eye.

There are four defects of human eye. One is myopia in which a person cannot see distant objects clearly. It is corrected by concave lenses. In hypermetropia, near objects are not clearly visible. This defect is corrected by convex lenses. In presbyopia, a person is not able to see both near and distant objects clearly. It is corrected by both

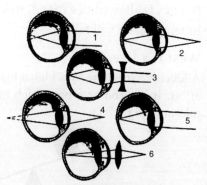

Fig. 27.2 Different types of defects of human eye

concave and convex lenses. In astigmatism, both vertical and horizontal lines are not focussed properly. It is corrected by using cylindrical lenses. (Fig. 27.2)

What are the properties of electrons?

Electrons are negatively charged particles and found outside the nucleus of an atom. The charge on the electron is 1.602×10^{-19} coulombs. Its mass 9.1×10^{-31} kg.

What is the speed of light?

Light travels with a speed of 300000 km/sec.

Name the three primary colours.

The primary colours are blue, green and red, which on mixing, produce white and other colours.

What is schlieren photography?

It is a photographic technique that reveals flow patterns in fluids by recording changes in refractive index.

What is a telephoto lens?

A camera lens of large focal length and narrow angle of view used to produce magnified image is called a telephoto lens.

What are Fraunhofer lines?

Dark lines in the spectrum of sunlight, named after Joseph von Fraunhofer, who first studied them. They are absorption lines, caused by the absorption of certain wavelengths by the gases in the sun's outer atmosphere.

What are the seven colours of white light?

Violet, indigo, blue, green, yellow, orange and red are the seven colours which form white light. (Fig. 27.3)

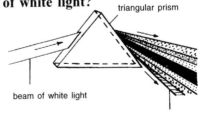

Fig. 27.3 Spectrum

What is interference?
When two waves of the same frequency travelling in the same direction are superposed on each other, they produce maxima and minima. This is known as interference of waves.

What type of mirrors are used in cars and buses for rear view?
Convex mirrors, because they have a wide field of view.

What should be the size of a plane mirror to see the full image of an object?
The size of a plane mirror should be half the size of an object to see its full image.

Give one use of concave mirror.
Concave mirror is used for shaving because it produces the enlarged image of the face.

What is phosphorescence?
The emission of light by a substance when it has been exposed to radiation. The substance continues to glow after the source of radiation has been removed. If the luminescence persists for a very short duration, its is called fluorescence.

What are the different parts of electromagnetic spectrum?
An electromagnetic spectrum consists of gamma rays, X-rays, ultraviolet rays, visible light, infrared rays, microwaves and radiowaves. (Fig. 27.4)

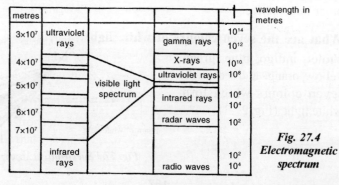

Fig. 27.4 Electromagnetic spectrum

What are microwaves?

Electromagnetic waves with wavelengths ranging from 1 mm to 30 cm. They are used in radar and in induction heating.

What are the two types of waves?

Two types of waves are transverse and longitudinal waves. Transverse waves are those in which particles vibrate in a direction perpendicular to the direction of wave propagation. In longitudinal waves particles vibrate in the direction of wave propagation. (Fig. 27.5)

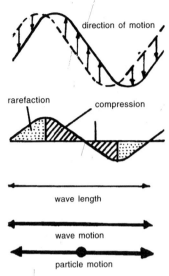

Fig. 27.5 Two types of waves

What are light waves?

Light waves are transverse electromagnetic waves.

What are sound waves?

Sound waves are longitudinal mechanical waves.

What are ultrasonic and infrasonic waves?

Man can hear sound waves of frequencies between 20 and 20,000 Hz. Frequencies above 20,000 Hz are called ultrasonic waves and below 20 Hz are called infrasonic waves.

What are shock waves?

Shock waves are the high pressure waves which are produced by explosions, earthquakes, volcanic erruptions, or by the movement of a body with very high velocity in air, gas or liquid.

How is an echo produced?

An echo is produced by the reflection of sound waves. When a sound wave gets reflected from some obstacle, it resounds as an echo.

Which element has the smallest atom?
Helium atom.

What is ball mill?
It is a machine that grinds lumps into powder consisting of a rotating drum containing pebbles or steel balls.

What is oxidation?
Loss of electrons is called oxidation.

What is reduction?
Gain of electrons is called reduction.

How do we define electrical resistivity?
The resistivity or specific resistance is the resistance offered by a conductor of unit length and unit area of cross-section.

What is meant by calorific value of a fuel?
The calorific value of a fuel is the quantity of heat which is produced when its unit mass is burnt.

What is the weight of a freely falling body?
A freely falling body becomes weightless.

How do we define mass?
The measure of inertia of a body is the measure of its mass.

What is weight?
The weight of an object is the force by which the body is attracted towards the centre of earth.

What are elasticity and plasticity?
When a force is applied on a solid it gets deformed but on removal of the force if the substance regains its original position it is said to be elastic and this property is called elasticity. However, if the solid does not regain its original shape on removal of the force, it is said to be plastic and this property is called plasticity.

What is electric current?
The rate of flow of charge is known as electric current.

Is water a simple compound of hydrogen and oxygen?

No, water is not a simple compound of oxygen and hydrogen. Oxygen has two isotopes, whereas hydrogen has three. Therefore, it is a complex compound of these isotopes.

Why does water rise in a capillary?

Water rises in a capillary due to surface tension.

What is a crystal?

If the atoms or molecules of a substance are arranged in a regular periodic fashion it is said to be crystalline material.

How are batteries connected to get more voltage?

The positive terminal of one battery is connected to the negative terminal of the other and so on. The voltage gets added up by this method.

What is quicksand?

An area of soft or loose wet sand of considerable depth, yielding under weight and hence apt to engulf persons, animals and objects coming on it.

What is a magnet?

It is a piece of iron or steel magnetised by special method. It acquires the property of attracting pieces of iron, cobalt or nickel. It is due to the magnetic field which surrounds the poles of a magnet.

What is a magnetic storm?

A magnetic storm is a disturbance in the magnetic field of earth. These are caused mainly by solar flares.

What is a magnetometer?

It measures magnetic forces.

What is the meaning of automation?

Automation means the use of some mechanism for the operation of machines without human help.

●●●

28. RECENT SCIENTIFIC ACHIEVEMENTS

What is a non-burning hot plate?

This hot plate was invented by a German engineer Anton Seelig in 1984. It can heat the contents of a saucepan without becoming hot itself. Heat is produced by electromagnetic induction, therefore children can place their hands on the surfaces of hot plate without any fear of being burnt.

What is the potato-powered clock?

The American company Skilkraft made a clock in 1984 powered by two potatoes. They provide power through two electrodes. The electrodes, one of zinc and the other of copper, inserted into the potatoes, induce a chemical reaction that generates electricity (Fig. 28.1)

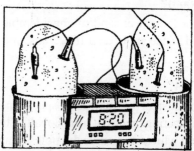

Fig. 28.1 Potato-powered clock

You have seen a keyboard typewriter. Can you think of a typewriter which types without a keyboard?

Japan developed a revolutionary typewriter in 1984 which does not have any keyboard. It has been named as Panaword. This typewriter has a tactile panel on which the characters, numbers and diverse signs are printed. The user writes his message on a tactile screen by hand with a special pencil and presses a button. He receives a typewritten text directly from the machine. The Panaword's memory recognizes 3448 characters.

What is a television watch?

A watch with a television is a recent development in which a small television receiver is fitted in the wrist watch (Fig. 28.2)

Fig. 28.2 Television watch

When was pocket television made?

Pocket television was made by Japan in 1984. The first black and white pocket television was the Sony Watchman which had a 5 cm diagonal screen. In 1985 Casio and other manufacturers designed many tiny sets in colour.

What is a flat screen television?

A flat screen television is a type of TV which can be hung on the wall just like a painting. The first operational model was designed in 1985 by a Japanese company Matsushita using liquid crystal technology.

Who was the first baby to be born from frozen embryo?

Zoe was the first infant to be born from a frozen embryo in Australia in the spring of 1984.

What is Karmarkar algorithm?

The Karmarkar algorithm, named after a 28-year-old Indian mathematician Narendra Karmarkar, is a remarkable breakthrough in the resolution of systems of equation which are often too complex for the most powerful computers. The general idea behind this is to bend the solid repeatedly and more rapidly to the optimal solution. American Petroleum and Aeronautics Company is working with Karmarkar algorithm.

What is the new definition of meter?

The new definition of meter adopted on October 20, 1983 is as follows: The meter is the length of the path covered by light in a vacuum in $1/299,792,458$ part of a second.

What is a microprocessor?
A microprocessor is a whole computer laid out on a surface of several square millimeter silicon chip. The circuit in the intel 4004 microprocessor contains 2300 transistors.

When was India's surface-to-air missile test-fired?
The surface-to-air missile 'Trishul' with a range upto 9 km., was successfully test-fired on December 28, 1996 from Chandipur near Balasore, Orissa. India has developed NAG, AKASH and Prithvi also.

When was the bubble memory designed?
The bubble memory was designed in 1967 by a team at Bell Laboratories (U.S.A.) under the leadership of Mr. A.H. Bobeck. It is a device whose contents are never lost even when power supplies are removed. Today, they can store 92 kilobits.

What is CCD memory?
The word CCD stands for Charged Coupled Devices. They are semiconductor versions of magnetic bubbles. They are mostly used for image sensing.

What is artificial intelligence?
Artificial intelligence is a scientific discipline consisting of writing computer programs that attempt to copy human intelligence.

What is a silicon chip?
A microchip is a silicon wafer which is chemically etched and contains distinct integrated circuits on it. These chips are used in computers and microprocessors. (Fig. 28.3)

Who isolated the AIDS virus first?
The team of Professor Montagnier of the Pasteur Foundation isolated the AIDS virus for the first time in January, 1983.

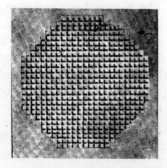

Fig. 28.3 Silicon chip

What is the latest revolution in robotics?

In 1984, Australian researchers have designed a robot which is capable of shearing a sheep. The shape of the sheep is stored in the computer's memory, which allows it to control the movements of the arm holding the electrical shear.

What is ELISA technique?

It is a technique which is used to detect the presence of AIDS virus in body fluids. Its kits are available almost in all countries of the world.

What is electronic publishing?

Electronic publishing is actually a catch all phase which covers many activities such as on-line databases, videotext, cable TV programming, teletext, videotape cassettes, videodisks, electronic mail, internet and messaging.

Which is the smallest word processor?

The world's smallest word processor is the Easitext 1350. It was introduced by Minimicro of Huntington, Yorkshire in April 1976. The entire system fits into an executive briefcase.

How does solar-powered street lighting system work?

A solar-powered street lighting system consists of a circular solar energy collector, the lamp, the pole and lead acid storage batteries. In the day time batteries are charged by solar energy and in the night time this stored energy is used for lamp lighting. The two batteries can hold a maximum of 100 Ampere-hours (Fig. 28.4)

Fig. 28.4 Solar-powered street light

Which is the world's fastest and most powerful computer?

World's most powerful and fastest computer is the liquid cooled CRAY-2.

Its memory has a capacity of 256 million 64 bit words, resulting in a capacity of 32 million bytes of main memory. It attains speed of 250 million floating point operation per second.

Where is the largest nuclear reactor situated?

World's largest nuclear reactor is the 1450 MW power reactor at the Ignalina Station, Lithuania. It was operated on full power in January, 1984.

For what purpose is CCD used in a telescope?

The attachment of an electronic charge coupled device (CCD) in a telescope increases its 'light grasp' by a factor upto 100 fold.

Which is the most powerful laser beam?

The most powerful laser beam is the 'Shiva' laser. It was reported to be concentrating 2.6×10^{13} watts into a pinhead-sized target for 9.5×10^{-11} sec. at the Lawrence Livermore Laboratory in a test on May 18, 1978. This laser is used for the activation of H_2 bomb which was first activated by atom bomb.

Which is the brightest light?

The brightest steady artificial light source is the laser beam. Of continuously burning sources, the most powerful is a 313 KW high pressure argon arc lamp. It was completed in March, 1984 by a Canadian Industry and is of 1,200,000 candle-power.

What is an acoustic microscope?

An acoustic microscope is a very recent invention of science. It uses very short wavelength sound waves focussed in super cooled liquid helium at 0.2 K to image samples. Its resolution is better than an optical microscope and is beginning to compare with the resolution of electron microscope. (Fig. 28.5)

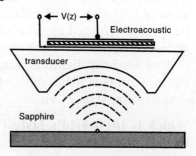

Fig. 28.5 Acoustic microscope

What does the term MRI stand for?

The term MRI stands for Magnetic Resonance Imaging. This is a method in which images of magnetic resonance phenomenon are used to detect the body diseases. A tumour of 1mm size can be detected by MRI.

Can a car be driven by simple voice commands?

Yes. The French scientists have developed a car with a mini computer which can be controlled by voice. For example, the words 'to the right' sets off the right turn signal.

What is a graphite fibre violin?

Mostly metal wires are used in violin but in 1983 Leonard Keith John of Canada made a violin using graphite fibres.

What is NMR scanner?

It is the most recent invention. NMR stands for Nuclear Magnetic Resonance. This is an instrument used to detect the body diseases. It is the most powerful tool for identification of diseases. It does not use any harmful radiations and gives better results than the CAT scanner. (Fig. 28.6). Delhi, Mumbai, Chennai, Kolkata etc. have NMR facilities.

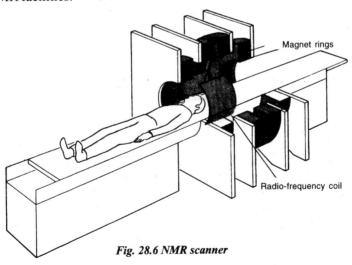

Fig. 28.6 NMR scanner

What is the recent development in Aerospace industry?
The Dupant Aerospace Company of USA has developed a prototype plane DP-2 which is able to take off and land vertically under light loads.

How are kidney stones broken down by shock waves?
Three doctors of Germany developed in 1982 an apparatus able to disintegrate a kidney stone of 3 cm in diameter using shockwaves. During the period of treatment which lasts for 1/2 to 3/4 hour, the patient receives about 500 shock waves, felt as a sort of slap. The stone is broken down into granules, which are then eliminated in the urine.

Who invented the mobile solar power plant?
The first wheeled solar power plant in the world was invented by an American Ty. Braswell in the early 1980s. It can be taken from one place to another.

Which is the largest infrared telescope?
The largest infrared telescope in the world, equipped with a lens having a 3.8 m aperture, was installed in 1977 by Great Britain at an altitude of 4200 meters at the top of Mount Mauna Kea.

When was the latest space craft launched for Mars probe?
The NASA launched a robotic roving vehicle called Spirit which landed on Mars on 4th Jan. 1904. It has started sending images of Mars.

Which is the hottest flame?
The hottest flame that can be produced is from carbon subnitride which at once can reach to a temperature of 5261 K.

Which is the first double quasar and when was it observed?
The first double quasar is 0957 + 56. Its existence was announced in May, 1980.

When was the longest telephone call made?
A telephone call around the world, over an estimated distance of

158845 km was made on 28th Dec. 1985 from and back to the Royal Institution, London, during one of the Christmas lectures given by David Pye. Multiple satellites were used for this call.

What is the highest vacuum that has been achieved?
The highest vacuum attained is of the order of 10^{-14} torr.

Which is the smallest hole which has been drilled so far?
The smallest hole size is 40 Angstrom (40×10^{-8} cm)

Which is the most powerful microscope?
The world's most powerful microscope is the scanning tunnelling microscope invented at the IBM Zurich Research Laboratory in 1981. It can magnify an object to 100 million times.

What is paging?
Paging is a one-way wireless communication providing instant access to the person being paged. The message is received by a small palmsize device called pager kept by the person with himself. The alarm sound from the device alerts the person who then reads the message flashed on the screen of the pager.

What are Kangaroo Shoes?
These are the special type of shoes developed in 1982 by a Canadian, David Lekhmann, which allow one to run using giant steps.

Is there any evidence of water inside a Lunar crater?
On December 4, 1996, it was said by Pentagon that the unmanned clementine spacecraft located a small mass of ice in a giant crater near the south pole of the moon, which had long been thought to be dry one. The discovery of water on the moon, perhaps deposited by a commet, increases chances that man may some day be able to take up residence on Lunar surface.

Who first invented a 360-degree view Camera?
A young Indian scientist Shree Nair of Columbia University, has constructed a complete little prototype of a camera that would yield a 360-degree view of the scene focussed on. It was invented in 1996.

What is an optical fibre?
An optical fibre is a thin strand of glass or plastic made from two types of materials. The inner part of a fibre is called core while the outer cover is called clad. It can carry light for long distances with the help of total internal reflection.

Who invented optical fibres?
The practical optical fibres were developed in 1955 by an Indian scientist Dr. Narinder S. Kapany working at Imperial College, London.

What are the uses of optical fibres?
Optical fibres are used in endoscopes to carry light to the stomach of a patient. They are being used for optical communications. They can carry large amount of information.

What is the future of optical fibres?
Since cables of optical fibres can carry large number of messages, they can be used to link homes to videotext commuters, video-telephones and facsimile transmission services.

What is LDR?
The term LDR stands for Light Detecting Resistance. It is made of cadmium sulphide. When light falls on LDR, photo electrons are emitted and the current flows in the circuit.

What is thermister?
It is a device used to measure small changes in temperature. These devices are made of silicon carbide, nickel and manganese oxide.

What is a hypodermic syringe?
Hypodermic syringe is used to inject vaccines and drugs. Hypodermic means 'below the skin'. This is a hollow needle.

●●●

29. SCIENTIFIC ABBREVIATIONS

AA	–	Anti-aircraft
ABM	–	Anti-Ballistic Missile
AC	–	Alternating Current
AF	–	Audio Frequency
AM	–	Ante-meridiem (Before noon)
ASLV	–	Augmented Satellite Launch Vehicle
ATS	–	Anti-Tetanus Serum
BCG	–	Bacillus Calmette Guerin-anti tuberculosis vaccine.
BP	–	Blood Pressure, Boiling Point
B. Th. U.	–	British Thermal Unit
C	–	Centigrade
COMSAT	–	Communication Satellite
CTV	–	Colour Television
CVR	–	Cokpit Voice Recorder
CW	–	Chemical Warfare
CWt	–	Hundredweight (112 lbs)
DC	–	Direct Current
DDT	–	Dichloro-Diphenyl Trichloroethane
DNA	–	Deoxyribonucleic acid
ECG	–	Electro-Cardio-Gram
ECT	–	Electro-Convulsion Therapy (electric shock treatment)
EEG	–	Electro-encephalogram
EMF	–	Electro-motive Force
EMU	–	Electro Multiple Unit

EVR	– Electro Video Recording	LCA	– Light Combat Aircraft
FBTR	– Fast Breeder Test Reactor	LES	– Lunar Escape System
FPS	– Foot-Pound-Second (Units of measurement)	LM	– Lunar Module
		LMG	– Light Machine Gun
GMT	– Greenwich Mean Time	LPG	– Liquefied Petroleum Gas
Hi-Fi	– High Fidelity	LSD	– Lysergic Acid Diethylamide
HE	– High Explosive		
HMT	– Hand Micro Telephone	MHD	– Magneto-hydro Dynamics
HP	– Horse Power	MOx	– Mixed Oxide Fuel
HSD	– High Speed Diesel	MPG	– Miles Per Gallon
Kg	– Kilogramme	MPH	– Miles Per Hour
KVA	– Kilo-volt-Ampere	MTB	– Motor Torpedo Boat
KW	– Kilowatt	NFC	– Nuclear Fuel Complex
Laser	– Light amplification by stimulated emission of radiation	NTP	– Normal Temeprature and Pressure
		NTPC	– National Thermal Power Corporation
ICBM	– Inter-Continental Ballistic Missile		
IRBM	– Intermediate Range Ballistic Missile	OTEC	– Ocean Thermal Energy Conversion

OZ	–	Ounce, Ounces
PM	–	Post Mortem (after death)
PSLV	–	Polar Satellite Launch Vehicle
RADAR	–	Radio Detection and Ranging
RNA	–	Ribonucleic Acid
ROBERT	–	Rocket Borne Emergency Radio Transmitter
RPM	–	Revolutions Per Minute
SAM	–	Surface-to-Air-Missile
SHM	–	Simple Harmonic Motion
SLBM	–	Seal-Launch-Ballistic Missile
SLV	–	Satellite Launch Vehicle
SRAM	–	Short-Range Attack Missile
S.S.	–	Steam Ship
STP	–	Standard Temperature and Pressure
TAB	–	Tetanus Anti Bacilli
TB	–	Tuberculosis
TELEX	–	Teleprinter Exchange
THI	–	Temperature Humidity Index
TNT	–	Tri-nitro-toluene
TSE	–	Total Solar Eclipse
UHF	–	Ultra High Frequency
UFO	–	Unidentified Flying Object
VHF	–	Very High Frequency
VTOL	–	Vertical take-off and Landing
ZPG	–	Zero Population Growth

●●●

Author: Manasvi Vohra
Format: Paperback
Language: English
Pages: 184

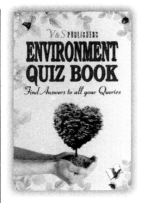

Author: Manasvi Vohra
Format: Paperback
Language: English
Pages: 144

Author: Vikas Khatri
Format: Paperback
Language: English
Pages: 152

History is something that occurred in the past which is memorable, or remarkable, good or bad. The word, history is basically used for a record of events that happened in the past.

In the book, Quiz Time History, there are about 1100 interesting, knowledge-based questions with answers that will educate the readers about the significant historical facts and incidents that took place during the Early Indian Historical Period, The Medieval Times and the Modern Era.

India, as we all know is a vast country with a rich culture, heritage and historical background.

The study of environment is important for us as we are an integral part of the environment. It includes composite physical and biological sciences including subjects, such as Ecology, Botany, Zoology, Physics, Chemistry, Soil Science, Geography, etc. Environmental studies also incorporate human relationships, perceptions and policies towards the environment. Hence, in order to understand and learn more about the environment; and to find answers queries people consider mysteries nature, Environment Quiz Book is an ideal one.

The book includes several interesting and simple:
- Questions & Answers
- MCQs • Fill in the Blanks
- Crossword •Word Search
- True & False

- Enjoy mental workouts?
- Like numerical brain teasers?
- Dabble in solving puzzles?
- Use maths occasionally?
- Accept intellectual challenges?
- Love solving riddles?

It "YES" to any of these questions, then this is the right book for you! Also if you want to test your logical skills and also to have fun, then read this collection of brain teasers and check out how smart you are!!

visit our online bookstore: www.vspublishers.com

Also available in Hindi

Author: Vikas Khatri
Format: Paperback
Language: English
Pages: 120

Author: Prosenjit Saha
Format: Paperback
Language: English
Pages: 108

Author: A.H. Hashmi
Format: Paperback
Language: English
Pages: 122

54 cool and Challenging art working, projects, crafts, experiments and more for kids!!!

Unplugging kids from their MP3 players and game systems for one-on-one family time is a great way to reconnect in today's hectic world. And what better way to spend time together than doing an activity that's not only fun but also promotes creativity and self-expression?

Greatest Crafts and Projects for Children is packed with 54 craft projects ranging from outdoor projects to gifts and party favours to holiday decor to projects that promote learning through play with step-by-step instructions to guide children to successful completion of each project.

We believe everyone can draw or paint. Of course some people are naturally talented but we are all capable of channelling our artistic skills and creativity.

With this belief in mind, we have published this Drawing and Painting Course Volume II for children who want to learn and master the art in a fun way. This book starts with the basics – lines, shades, texture, balance, harmony, rhythm, tone, colours, etc., and goes on to teach the various different techniques of drawing and painting with step-by-step instructions, accompanied by an audio-visual CD.

Children have always been attracted towards bright colours, various shapes and diverse objects that they see around them. Nature fascinates them. The beautiful birds, animals, flowers and trees fire their imagination and they want to capture it on paper. But how, for all are not artists by birth.

Well, this book has been especially developed for those who want to learn and master the art in a fun way. The step-by-step instructions, along with the audio-visual CD, will show you how to create beautiful pictures. See how a circle or an oval transforms into a flower or a peacock; a few lines here, and a few there become a human figure.

visit our online bookstore: www.vspublishers.com

Also available in Hindi, Bangla, Tamil

Author: Dr. C.L. Garg/ Amit Garg
Format: Paperback
Language: English
Pages: 120

Science projects and models play a pivotal role in inculcating scientific temper in young minds and in harnessing their skills. school students of senior classes have to work on such projects and these carry much weight in their overall performance.

All these aspects have been considered during the compilation of the projects and models. This book will also be an ideal choice for parents interested in enhancing scientific temper of their children; and for hobbyists.

The book has 81 Classroom projects on: Physics, Chemistry, Biology& Electronics for Sec. & Sr. Sec. Students

Also available in Hindi

Author: Vikas Khatri
Format: Paperback
Language: English
Pages: 160

A study of Science and Scientific theories is almost incomplete without relevant and methodical Experiments. In fact, experiments are an inseparable part of any scientific study or research. In this book, the author has tried to simplify science to the readers, particularly the school-going students, through easy and interesting experiments. The experiments given in the book are based on one scientific phenomena or another, such as atmospheric pressure, high and low temperatures, boiling, freezing and melting points of solids, liquids, gases, gravitational force, magnetism, electricity, solubility of substances, etc.

Also available in Hindi

Author: Ivar Utial
Format: Paperback
Language: English
Pages: 144

Supplementary science books not only interest and excite young students, but also stimulate their interest in the subject.

This exciting book shows you how to have fun with 101 Science Games. There is little doubt that science experiments can be quite interesting and useful in discovering mysteries of nature.

The book is fully illustrated with step-by-step instructions to give you first hand experience of making simple scientific equipments like:

Telescope
Barometer
Hectometer
Model Electric Motor
Electroscope
Periscope
Steam Turbine; and more...

visit our online bookstore: www.vspublishers.com

Author: Sumita Bose
Format: Paperback
Language: English
Pages: 212 (Fully colour)

Author: Ivar Utial
Format: Paperback
Language: English
Pages: 127

Author: Gladys Ambat
Format: Paperback
Language: English
Pages: 256

Learning Mathematics – The Fun Way caters to the students who deserve to have their individual learning needs satisfied. This book emphasises on teaching with activities, drawing on real-life models from children's point of view and promotes expectations for success.

The book nurtures the interest of the student by bringing up the fun-quotient in the learning process. This will help them gain confidence in ability to reason and thereby increase problem solving skills. The book also meets the requirement of the latest NCERT syllabus.

Enliven your leisure hours with QuizTime! It guarantees you to give many hours of exciting mind storming quiz games. Excel your ability to hold social meetings with charisma and quiz gaming. This book employs tested quiz skills in very well-defined structure for easy comprehension. The book is aimed to cater to a large section of the society.

Quiz and puzzles are brain fitness fundas of a unique kind! The thrill to win or lose gaming session of a quiz programme can give you an optimum level of mental fitness and alertness. You simply bubble over with the sheer joy of challenge.

The book is a lively presentation for all youngsters and a pleasant leisure companion for the elders. The veteran author has put together over 4000 exciting quizzes and interesting brain-teasers to get you all keyed up. While you race through every page — you could find yourself sitting on the edge of the chair. Yet, you get charged with a spirit of challenge to unearth hidden answers or solve uncharted problems by your latent thinking power.

visit our online bookstore: www.vspublishers.com